# MAN AND THE MARITIME ENVIRONMENT

*Edited by*

**STEPHEN FISHER**

UNIVERSITY
*of*
EXETER
PRESS

First published in 1994 by
University of Exeter Press
Reed Hall
Streatham Drive
Exeter EX4 4QR
UK

**British Library Cataloguing in Publication Data**

A catalogue record of this book is available from the British Library

ISBN 0 85989 393 6

Typeset in Palatino by Sue Milward

Printed in Great Britain by Short Run Press Ltd.

# MAN AND THE MARITIME ENVIRONMENT

# EXETER MARITIME STUDIES

*General Editor :* Stephen Fisher

To all my shipmates, past and present,
for adding to my understanding
(The Editor)

# ACKNOWLEDGEMENTS

The editor of this collection wishes to thank Jane Ashby for her estimable skill in word processing, Sue Milward for her professional expertise in typesetting, and Rodney Fry for assistance with the Figures. Delphine Jones of the University's Graphic Design Unit kindly designed the book's cover. Simon Baker, the Secretary of the University Press, has been an indefatigable and resourceful collaborator. Special thanks are appropriate to a benefactor who wishes to remain anonymous for meeting part of the pre-publication costs. And, finally, much gratitude is owed to the members of the Dartington conferences - a changing group year by year - for making our meetings such congenial and worthwhile occasions.

# Contents

# List of Plates

Front Cover:     Plundering the rock pools at Ilfracombe, North Devon (1860s).
(Reproduced by kind permission of Ilfracombe Museum)

# List of Figures

# INTRODUCTION

In recent years the annual Dartington maritime history conferences organised by Exeter University have broadened their earlier interest in mainly South West of England issues to assume a more national and international character. This broadening is reflected in the papers below emanating from the conferences held in October 1991 and 1992. They focus on aspects of man's involvement with the sea as an environment, in particular as marine scientists, as seafarers, and in the exploitation of the sea and the seaside for recreation. It is hoped that these papers will make some contribution towards our developing understanding of man and the maritime environment, an aspect of 'environmental studies' which is now being keenly advocated.[1]

Four papers deal with man's scientific interest in the sea, initially with the rise of marine science in Britain and the United States, and then with the work and findings of two research institutions in England, the Cornish Biological Records Unit at Truro and the Laboratory of the Marine Biological Association at Plymouth.

Thus Margaret Deacon (herself the daughter of an oceanographer, William Deacon) surveys the gradual growth of British government support for oceanographic research from the mid-nineteenth century on to the setting up in 1949 of the National Institute of Oceanography (now the Institute of Oceanographic Sciences). Nowadays fundamental research, it is generally accepted, needs public support, although which areas of knowledge and to what extent provides a fertile ground for debate. But in the mid-nineteenth century the situation was quite different, the prevailing 'laissez faire' attitudes on the government's role and public expenditure applying also to fundamental research. By its very nature oceanographic research was particularly demanding of resources, little being capable of being done without expensive ships and manpower. Gradually from the mid-century government reluctance over the financing of marine science changed, stemming essentially from the manifest incapacity of private resources to deal with the pressure of practical problems. Two issues in the 1850s were critical: first, the need for knowledge of the seabed to facilitate the efficient laying of underwater telegraphic cables, and secondly, increasing international interest, following a Brussels conference in 1853, in the systematic observation of weather at sea, this interest leading to the setting up of the

1

Meteorological Department of the Board of Trade (the forerunner of the present-day Meteorological Office). A later, major step forward was the public financing of the *Challenger* circumnavigation of 1872-6, to explore the influence of temperature on the circulation of ocean water masses. This expedition has been seen by many as the starting point of modern marine science, since it led scientists working on different aspects of oceanography to see themselves as working within a common discipline. Dr Deacon offers a fascinating discussion of the later evolution of public attitudes, and the continuing influence of 'deserving projects' including the interest of the Royal Navy, such as the early twentieth-century concerns with the science of underwater acoustics following on the development of submarine warfare. Her paper ends with the sharp remark of Lyon Playfair in the 1880s that it was 'really disgraceful' that a nation which owed so much of its prosperity to the sea should have been so slow to develop fundamental knowledge of it.

The practical *raison d'être* for the state funding of marine scientific research also comes through strongly in Harry Scheiber's paper on modern Pacific oceanography, involving United States federal finance. This paper focuses on the burgeoning of Pacific fishery-related researches after World War II, involving cumulative funding of 'tens of millions of dollars and the fulltime employment of some ten vessels'. Scheiber has two main themes: first, to explore and establish that the ecosystem approach the Americans adopted was very much linked to the Northern European and British legacy of 'ecological' research developed from the 1880s. And then to examine the parallels between British and American experience in the 'critical moments' of new funding that spectacularly advance marine science (as a guide for the present perhaps as to how the vitality of ocean research can be furthered).

On this latter point Prof. Scheiber reveals some telling factors at work. In the United States prior to 1945 there was only limited government support for West Coast marine research. Scheiber identifies a number of imperatives for greatly expanded funding, strategic, political and the support of powerful economic groups. He discusses the Truman Fisheries Proclamation of September 1945, in which the United States abandoned the traditional 'three mile' interest to define 'conservation zones' far out to sea. Then the U.S. Navy gave 'enthusiastic support' in 1946-7 to oceanographic research that could have military importance. But the support of West Coast fishing interests was also compelling. By 1948 the California sardine industry, faced by a rapid drop in catches and fearing the disappearance of the fishery altogether, lined up with the scientists to press for government support of an ambitious research programme. The powerful West

Coast tuna fishing interests (whose operations covered the entire Eastern and Central Pacific regions) also supported other research projects. Scheiber has revealed the beginning of the movement that established oceanography as 'big science' in the United States.

An issue of both the Deacon and Scheiber papers is the role of thrusting entrepreneurial individuals (to be found in academe as elsewhere) in marshalling the arguments and wider support necessary to persuade public and private agencies to fund research. A nice instance of this concerning the universities themselves assuming financial responsibilities is the enterprise of two scientists in 1947 who persuaded a reluctant University of California president to acquire two refitted U.S. Navy vessels and a fishing vessel for Pacific deepwater research - as the president put it, the university acquired its own navy overnight.

Now to the papers from English marine biological institutes. Stella Maris Turk considers the work of the Cornish Biological Records Unit, which is funded by Exeter University and Cornwall County Council. Established in 1972, its purpose was and is to compile records of all the species of plants and animals that have lived or presently live in Cornwall and the Isles of Scilly, including the shores and adjacent seas, some 21,000 species. On the manual index, in the mid-1980s, there were over a million biological records, recognised as being the largest and most comprehensive regional data bank in Europe. A computer programme, ERICA, is being set up, and 400,000 records of 17,000 plants and animals are currently capable of access (Prime Computer UK and British Telecom have been generous sponsors of this facility).

The author further considers why Cornwall has such a rich diversity of marine life, the county being noted for its mix of 'northern' and 'southern' including 'Mediterranean' marine organisms. She also considers, using the Unit's resources, marine biological changes over time, sometimes long-term, and the factors that seem to be at work. They include man's intervention in the maritime environment in the form of tourism (Cornwall being a holidaymaker's as well as a naturalist's paradise) and industrial and other pollution. The most important event threatening West Cornwall's marine habitats was the major oil spill caused when the oil tanker, *Torrey Canyon*, hit the Seven Stones reef in 1967. Out of such disasters though can come progress, the lack of basic information to assess its effects contributing to the setting up of the Biological Records Unit itself.

In contrast Drs Alan Southward and Gerald Boalch present a very wide-ranging essay on changing world climate and marine life. Their initial concern rests on the sea's role as a controlling and moderating influence on global climate, and particularly with the evidence for past changes in nearby English Channel sea temperatures and consequent marine biological changes, as researched by themselves and their colleagues at the Plymouth Marine Biological Laboratory.

Serious observation of sea temperatures began in Britain in the 1860s when the Meteorological Office asked merchant vessels to record weather observations, although there are earlier American, non-sustained, series such as the Maury observations of the Gulf Stream. Our authors indicate the warming trend of sea temperatures in the western Channel and Bay of Biscay waters in the first half of the twentieth century, and the cooler trend that set in after 1961 until quite recently. Using various sources including the, initially at least irregular, historical records of fish catches and trade from the sixteenth century onwards, and the Plymouth Laboratory's more recent serial observations, they show that cold-water marine organisms have varied in abundance over time, and that after 1930, of the mid-water fishes, the herring shoals failed to maintain themselves (by 1936, the Plymouth herring fishery had collapsed) with the warmer-water pilchard becoming more abundant. Similar changes occurred in the sea-bottom living fishes. The post-1961 ending of the warming trend in sea temperatures saw these biological changes reversed, while the post-1980 reversal of sea-water cooling has again seen increases in warm-water life.

The implications of anthropogenic influences on recent (post late-1950s) global temperatures are then considered, that is the rising $CO_2$ and 'greenhouse' effects - the existence or not of significant-for-trend 'greenhouse warming' is judged to be a 'qualified perhaps'. Some predictions are made as to what will happen to climate and marine life in the next century (it is striking to a social scientist how prediction comes easily to the natural scientist - they, of course, say it is both logically inherent in their work and necessary). If sea temperatures rise further then the increase in southern warm-water species detected at Plymouth since 1980 will continue and 'may go beyond what was observed in the previous peak years of warming from 1940 to 1960'. In conclusion, Southward and Boalch (rather reassuringly) caution against scenarios that include 'runaway' effects, pointing to the negative feedbacks that must exist in the global carbon cycle, evident from the past record of rocks and sediments - 'ultimately the ecosystem has the ability to cope with increases in global carbon dioxide and achieve stability' (but note the 'ultimately'). They end by suggesting that the recent global warming trend has been superimposed on a more 'natural' oscillation of the system, possibly related to medium and long-term cycles in solar activity. Where 'so many overlapping factors of importance' coexist, detailed data series can be crucial for satisfactory trend analysis. Thus their plea, that the suspension of marine monitoring at the Plymouth Laboratory after 1988 on grounds of economy needs to be reversed.

As an historian one is struck both by the cogency of the analysis and the likely value of resumed Plymouth monitoring. One is also impressed by the part historical evidence familiar to many economic historians plays in such analyses (and strikingly so in the historical climate studies of H.H. Lamb[2] and the University of East Anglia School of Historical Climatology). Far more inter-disciplinary collaboration between biologists and historians seems to be called for, in these critical areas of investigation into the marine environment and human influences.

Michael Stammers, Keeper of the Merseyside Maritime Museum, has written an unusual, indeed pioneering, appreciation essentially of the non-depredatory relationship of sailing-ship seafarers of the nineteenth and early twentieth centuries with the sea creatures they encountered. His sources are British, including professional journals such as the *Nautical Magazine*, biographies, diaries and the like.

Man's attitude to the sea has always been ambivalent - it can be a provider of riches and a means of communication, but it is also capricious and destructive. And ambivalent was his attitude to sea creatures. Seafarers seem to have had no affection as such for the creatures they observed. Some, such as sharks, were seen as outright enemies; others, like flying fish, were welcome supplements to everyday rations. But if not exciting the affection of seamen, who were often enough engaged in their slaughter, sea creatures could evoke wonder and admiration, most notably whales, dolphins, porpoises and seals, with the albatross, sometimes with a wingspan of over 11 feet, thought to be the most beautiful, the most awe-inspiring. Dr Stammers also briefly considers the interest of seafarers in formal natural history, beyond what professional seamanship demanded in the way of knowledge. Systematic study starts with the Royal Navy and Captain Cook in the eighteenth century. In 1832 the *Nautical Magazine* was enthusiastically urging naval and merchant officers to keep scientific records on their voyages. And in 1861 a Liverpool doctor, Cuthbert Collingwood, urged the ships' captains and shipowners of his native port to take up the 'hobby' of natural history, and the city's Literary and Philosophical Society published a handbook on how to collect and preserve specimens. Our author concludes that seafarers at this time in their attitudes to the creatures of their environment were no better and no worse than were landsmen in theirs. Dr Stammers has opened up an issue worthy of much fuller investigation, over both time and national boundaries.

Two papers then deal with the relationship of man with the coastal marine environment, in the rise of the seaside tourist industries and the apparently inevitable spoliation of that environment by their very success. Both focus on aspects of the experiences found within Devon, but serve as more widely representative case studies, certainly of that development of seaside tourism springing out of the natural attractions of sea and foreshore.

John Travis' contribution derives from his recently published investigation into the rise of Devon seaside tourism from its beginnings in the mid-eighteenth century to the close of the nineteenth century.[3] Here he shows how the various attractions of the Devon-coasts (which had produced by 1900 a burgeoning if almost entirely select and non-mass tourism) had in every case, and often quite quickly, become damaged - 'severely damaged' as he in fact expresses it.

Members of the leisured class were first encouraged to visit the English coasts by fashionable medical practitioners, in 1750 Dr Richard Russell, for instance, publishing a work which extolled the sea's health-giving effects. Russell prescribed not only sea bathing, which he saw as a medicated bath, but also the curative drinking of sea water - an extension of the long-established 'taking the waters' at inland spas. But with the visitors came their effluent and consequent pollution of the sea. By the late nineteenth century the leading resorts had embarked on ambitious schemes to extend sewer outfalls to points where 'tidal currents could carry away the noxious wastes', but with only limited success, and, as we know to our cost today, the problem still engages local authorities. The other Devonshire seaside attractions were fresh sea air, splendid coastal scenery and the rich diversity of animal and vegetable life on the sea-shore. In each case man is shown to have caused deterioration of that which he had come to enjoy, through the building and smoke pollution of town houses and the railway, and the pillaging of seaside marine biological specimens. Travis cites Charles Kingsley's evident distress at the unforeseen damage to the marine environment caused by his book *Westward Ho!*. While Edmund Gosse, whose father Philip and his followers had encouraged amateur naturalists to visit the Devon coast spoke of the 'calamity that [his father] had never anticipated' wrought by the 'army of collectors' that had roamed over the shores, and the 'great chagrin' it had cost him. Dr Travis also offers some thought-provoking observations on noise pollution in the later nineteenth century. Our author has written a revealing, well-researched account of man's early depredations on the marine environment he had come to marvel at, and raises issues which are still relevant.

John Channon offers a specific case study of the pollution problems in a notable south Devon resort, Teignmouth, and the steps that have been taken at different times in the twentieth century, in the

Edwardian period and especially at the end of the 1980s, to remedy it. He reiterates that issues of tourist-related pollution are far from new, being present in the decades before 1914. In Edwardian times by far the most important sanitary problems in Teignmouth were the pollution of the river Teign and the need to improve the resort's water supply. Cases of typhoid were reported after bathing in the river, and the eating of cockles and mussels taken from the river was frequently noted in health reports as hazardous. The water supply problems were solved by a scheme inaugurated in July 1908 which brought in plentiful fresh water from the uplands of Dartmoor. But the pollution of the resort's beaches caused by inadequate sewage disposal facilities was to be of longer standing, and the author examines the very recent large-scale project to remedy this begun in the later 1980s. The trigger for this may been seen as the campaigns, both national and following on from European Community directives, for cleaner beaches and inshore waters. He details the massive scheme involving the spending by South West Water (the responsible authority) of some £30 millions which should solve matters. Dr Channon also looks at other local issues, of 'visual' pollution and the possible environmental damage that some think has occurred or may occur from projected infra-structural and housing developments. The paper illustrates well the dilemmas facing resort authorities who both wish to maintain and improve the environments that attract tourists and their incomes, and to respond to the needs of others for non-tourist related employment and housing and better amenities.

The final contributions to this volume raise other and wider issues to do with man and the maritime environment, in that they focus on man as a seaman. Thus Kim Montin, a postgraduate researcher at the University of Åbo in Finland considers the issue of how seamen regard or conceive of themselves, their work and their environment. This study is part of a larger research project in mari-time ethnology being undertaken at Åbo. The paper can be seen per-haps as one of the 'first fruits' of a formally established collaboration between the maritime researchers of Åbo and Exeter Universities.[4]

Montin's study is comparative. He looks at Finnish seamen from three time-periods over this century, namely the final stage of the sail-ing ship era, steamers in the 1950s and 1960s, and motor-driven ves-sels at the beginning of the 1990s. His study is confined to Finnish cargo vessels on routes mainly in the North Sea and the Baltic, to pro-vide some continuity of comparison. Like all Scandinavian research-ers Montin is more theoretically driven than is common in the more empirical Anglo-Saxon world of historical research, in his case being inspired by Knut Weibust's interest in the miniature society of crews on sailing vessels, the key concepts being 'role', 'norm' and 'status'.

Montin makes use of a variety of sources, including literary evidence but also questionnaires and interview material and what might be called 'participatory observation'. He has followed Prof. Tawney's dictum, that researchers should 'use the boot as well as the book', in that he has lived aboard a modern 3286-ton Finnish lift-on lift-off vessel on a two-week voyage from Kotka in the Gulf of Finland to Scotland, the Thames, Rotterdam and back to Helsinki. As he says there is nothing like actual experience of this sort to understand what really goes on - an alternative tactic, of course, would be to work professionally on such a vessel, but as well as being impracticable, this could easily narrow the vision of the investigator. Montin discusses in a most interesting way the range of questions he seeks to answer, including what constitutes the 'representative seaman', using Richard Henry Dana's observations of 150 years ago and applying them to the modern sailor. He is especially concerned with the 'space' of the seaman and what factors affect how space is allocated to different purposes on a vessel. Altogether the author offers a challenging and stimulating set of thoughts on his research - and opens a further window on the diverse nature of what constitutes man and the maritime environment.

In the concluding paper, Alston Kennerley, both a certificated ship's master and, more latterly, an established lecturer and researcher in marine studies at the University of Plymouth, examines the theme of man and the maritime environment in terms of occupational conditioning, of the socialisation of man to a career at sea. By virtue of his two careers, the former with its practical sea experience and the latter yielding theoretical and literary insights into maritime affairs, Kennerley is especially well prepared to consider such issues. Moreover, his first two sea-career voyages, in 1952, were as a merchant navy officer-cadet on board the celebrated four-masted sailing barque, *Passat*, and his recollections of his time on board this vessel serve as first-hand evidence of socialisation and industrial learning.

Dr Kennerley commences by briefly reviewing the state of theoretical knowledge and offers some new empirical evidence on the question of the choice of the sea as a career. He has found that of the 194 youngsters who entered the Boys' Department of the South Shields Marine School between 1886 and 1900 about half could be described as coming from seafaring families. Such a clear association can also be seen in the other cases the author discusses. In his own case, Kennerley is of the opinion that past family associations with the sea led him to take it up as a career, and to his being accepted at the age of 16 by the shipping line of Alfred Holt and Company of Liverpool as an apprentice officer. He was then given the chance to undertake two voyages as a cadet from German ports to South

America and back on *Passat*. Holts, and in particular the central figure of Lawrence Holt, had been influenced by Kurt Hahn of Gordonstoun and 'outward-bound' training fame, so that certain cadets were able to avail themselves of this remarkable beginning to their careers. *Passat* was being used at that time as a German sail-training vessel and Kurt Hahn was keen to bring German and British cadets together as a means of breaking down post-war barriers. The paper presents some of Kennerley's recollections (aided by a diary he kept at the time) of 'industrial socialisation' on a windjammer (he and his fellow cadets, German and British, were best pleased when a propeller on the recently-fitted auxiliary engine broke off). Our author rather drily concludes that in his personal sea-socialisation on the *Passat* (ships as he earlier notes have been likened to closed or 'total' communities, like prisons or monasteries or boarding schools) he had been well prepared by his own earlier boarding-school experience.

All of the papers below, in the themes they engage, indicate interesting areas of and approaches to further research. We need to know more on the development of marine science, on how marine life and marine sites are affected by both climatic changes and man's activities, and how man relates to the sea. This volume's essays may be seen as pertinent and useful contributions to these areas of investigation.

When these research papers were presented at Dartington, there were indeed lively discussions that ensued. In part the vitality of inter-change reflected the quality of the papers. But in good part it stemmed too from the traditional unique mix of the Dartington conferences, of theoretical (in varying degree) academics and part-time researchers from different professional backgrounds, and of others with a varied, practical experience of the sea. The recently-established Centre for Maritime Historical Studies at Exeter, and the Department of Economic and Social History before it, have derived great satisfaction from organising these conferences over the years with their marvellous mix of associations with the sea. Along with its research projects, its research students, and its new, taught-postgraduate degree in maritime history, the Exeter Centre has every intention of maintaining this special conference tradition, this unique blend of 'professional socialisation'.

March 1994                                      Stephen Fisher

## NOTES

1 For one succinct and challenging statement of the need for more knowledge in the field of man, his technology and the marine world, see the review by David Assinder of *Case Studies in Oceanography and Marine Affairs* (Pergamon and Open University, 1991), in *The Times Higher Education Supplement*, 26 June 1992, p.25.

2 See, for example, H.H. Lamb, *Weather, Climate and Human Affairs* (London, 1988) and other works.

3 Originally a PhD thesis submitted to the University of Exeter in 1988, John Travis' general work is now published as, *The Rise of the Devon Seaside Resorts, 1750-1900* (University of Exeter Press, 1993).

4 The Centre for Maritime Historical Studies at the University of Exeter, created in 1990, has collaborative research agreements involving both projects and staff/student exchanges, with the Universities of Åbo, of Newfoundland and of East Carolina.

# BRITISH GOVERNMENTAL ATTITUDES TO MARINE SCIENCE

## Margaret Deacon

In the last quarter of the nineteenth century, when scientists in Europe and America were seeking financial support for oceanographic research, the British government was frequently cited as a model of generosity and foresight. This was because in 1872 it had despatched at public expense the first major expedition for the scientific study of the sea. Scientists had long been interested in the sea[1] but it was only after the *Challenger* expedition that the different aspects of its study were recognized as forming a separate and coherent entity, the science of oceanography. The *Challenger* scientists had an opportunity to investigate the ocean on a scale unprecedented before. Indeed the American scholar H.L. Burstyn has described the undertaking as 'unique in its own century and scarcely duplicated in magnitude in ours'.[2] On this showing it would appear that the British government was generous to science in general and oceanography in particular but this was not always the case. Though Britain was active in fisheries research in the late nineteenth century other aspects of marine science were often neglected. Having pioneered the science of oceanography with the *Challenger* expedition it was not for another 75 years that a British government acknowledged the need for a permanent establishment for research in this field, with the foundation of the National Institute of Oceanography in 1949. What lay behind this apparent inconsistency?

Nowadays governments of most shades of political opinion accept that at least some areas of scientific research have to be paid for out of public funds because of the scale of work and the expense involved and because the information sought is important nationally rather than of interest to any particular branch of industry. However in the nineteenth century this point of view was not generally held in Britain, though in several other European countries, notably France and Germany, state support for science was well established before 1900. Historians of science have shown that during the century from 1850 to 1950 British governments only moved slowly and often reluctantly in this direction.[3] Their uncertainty in accepting the need for intervention mirrored more widespread reservations felt by society as a whole, especially during the nineteenth century, about the

11

desirability of state interference in this, as in other areas of national life.  During the period dealt with in this paper governments only intervened in marine science when individual and private resources were manifestly unable to respond to the challenge of some problem of national concern, such as fisheries, or when the country's external prestige was involved.  Even then they might be reluctant to do so for the support of science was not generally seen as part of their duty, even when strong arguments could be made that this was in the national interest.  During the nineteenth century this position was common to most politicians, irrespective of their affiliation.  However, attitudes both of individuals and governments did undergo slow change and during the first half of the twentieth century support was more willingly given, though still usually directed at problems of economic importance.  These and the need for oceanographic expertise in World War II ultimately won the argument for fundamental research into the marine sciences.

Because scientific exploration of the sea was difficult for individuals its development was different in many ways from what might have been expected if purely scientific factors had been decisive.  On the other hand economic links, especially with fisheries, meant that important work was done in areas which might not otherwise have been given priority.  Lack of resources was always the chief difficulty faced by scientists hoping to study the sea.  Compared with many other branches of science oceanography was demanding of resources, even in its early stages, because little could be achieved without ships and manpower, always expensive.  Areas where individuals could make significant contributions were therefore limited.  Partly for this reason it was not until the latter part of the nineteenth century that the different branches of marine science came to be seen as aspects of a single discipline, the science of oceanography.

This difficulty had not prevented scientists from being aware of the interest and importance of the science of the sea and from trying to do something about it at an early stage.  In 1633 the mathematician William Oughtred told seafarers to observe the 'diversities and seasons of the windes, and the secret motions or agitations of the Seas' to provide information that would enable scientists to improve navigation and safety at sea.[4]  Thirty years later the Royal Society issued 'Directions for Seamen', urging sailors to record details of tides, currents and depths of the sea, variation of the compass, weather, ship's position and similar details, information, as Henry Oldenburg put it, 'of good use, both *Naval* and *Philosophical*'.[5]

An Italian contemporary, Count Marsigli, described the problems he had encountered in similar attempts to observe depth, temperature, salinity and movements of the sea off the coast of Provence:

> No one could have foreseen the many great difficulties which the nature of the sea presents to those who wish to penetrate its interior and it is perhaps because of this that a number of intelligent men have not only been unable to see through a comparable exercise but have not even been able to start. The Public which is ordinarily interested only in its own affairs, viz. human existence, is content with a few facts relating to Navigation or to help distinguish between different species and qualities of edible fish ....[6]

Widespread indifference, technological difficulties and problems of access meant that scientists made slow progress in the investigation of the ocean over the next 150 years.

Marsigli concluded that further advances in this field would not be made until the state was ready to make resources available.[7] There were already precedents for royal initiatives in the support of science, as in the establishment of astronomical observatories, at Greenwich and elsewhere in Europe. One of the founders of the Royal Society, Sir Robert Moray appealed, unsuccessfully, for an observatory to be set up to study tides.[8] However, though some way from achieving Marsigli's ideal of the generous prince, British governments and scientists enjoyed a fruitful relationship in the realm of maritime science during the eighteenth and early nineteenth centuries, arguably more so than during the next hundred years.

This was because their aims had much in common. This can be seen in the arguments used when the Royal Society, and individual scientists like the astronomer Thomas Hornsby, calling on the government to send expeditions to observe the transit of Venus in 1761 and 1769, 'stressed ... the competitive and nationalistic aspect of the undertaking', as well as wider economic considerations.[9] Scientists viewed the collection of scientific data as well as plants and other products from newly explored lands as much in terms of the benefits to be expected for society as a whole as for the gain of knowledge for its own sake. In their turn governments had powerful political, economic and social motives for exploration[10] and were prepared to encourage science in related fields through actions such as, in 1714, founding the Board of Longitude to encourage and reward the discovery of methods of determining longitude at sea. It was these that would make possible the great age of scientific maritime exploration, from Cook onwards, in the latter part of the eighteenth century. Many years earlier Edward Halley's voyage to the South Atlantic in 1699-1700 had been undertaken principally to observe magnetic variation in the hope that this could be used to calculate longitude.[11] The culmination of this trend was a close relationship between government and scientists, such as David Mackay has shown in the case of Sir Joseph Banks.[12] Another influential voice in the decades immediately before and after 1800 was that of the geographer James Rennell.

Accurate navigation meant that it was possible to collect information about sea surface conditions and weather on a systematic basis. This kind of material enabled scientists and geographers to learn more about world climate, the course of ocean currents etc. To extend such observations to the ocean depths however was a different matter, involving delays, physical labour, and risk of damage or loss to irreplaceable apparatus. It is really surprising how much, rather than how little, was attempted in such conditions, but usually only when the captain himself had a direct personal interest in the proceedings. On Cook's voyages only a few subsurface temperature measurements are recorded but the contemporary Arctic voyage of Constantine Phipps, though turned back by pack ice, yielded several deep-sea soundings, as well as other interesting observations.[13] Similar work was carried out on several of the voyages in search of the North-West Passage sent out by the Admiralty from 1818 onwards. Here too the most interesting observations from the oceanographer's point of view were made on the least successful of the expeditions geographically speaking, that commanded by Sir John Ross in 1818.[14] At this stage there were close contacts between interested seafarers and scientists like Alexander Marcet who was interested in the distribution of salinity in the ocean and James Rennell who charted the currents of the Atlantic and other oceans.[15] Nevertheless sailors faced problems of interpretation as well as technical difficulties and these increased rather than diminished as time went on. This may be seen particularly in the otherwise impressive deep-sea temperatures and soundings made by Sir James Clark Ross during his Antarctic voyage of 1839-43.[16] Unfortunately he did not realize that his results were being vitiated at greater depths by shortcomings in his thermometers. Better communication between different branches of science could have alerted him to this danger but his difficulties were compounded by the fact that interest in these areas of marine science had waned in mid-nineteenth century Britain. At this period her scientists were pre-eminent in the exploration of marine life of the seashore and coastal waters, an activity well-suited to the amateur.

In the early nineteenth century British science was largely an individual and amateur activity. A few practitioners had university posts, a few held government appointments, like the Astronomer Royal, and a few were supported by private foundations such as the Royal Institution. Many important discoveries were made by people working independently, supporting themselves either by private means or through a profession. Medicine and biology were closely linked since medical training attracted many with scientific leanings. This gave rise to the need for greater collaboration and contact between individuals and this led to the formation of specialized scientific societies during this period. However while performing an important intellectual function these did little to help with the material needs of scientific research. Realizing the need for help of a more practical nature scientists joined together at the beginning of the 1830s to found the British Association for the Advancement of Science. This organization distributed funds raised from members via a system of committees. It supported many valuable projects, including a programme of dredging in British seas,[17] but never had sufficient funds at its disposal to establish or maintain permanent scientific institutions. The problem of science as regards the work of individuals was partially recognized and addressed by the government which in 1849 offered an annual sum to the Royal Society to be used for grants to support individual research. In exceptional cases, where a specific public interest had been identified, the government itself might even take scientists into its employ, as for example in the Geological Survey, established in 1835. Its work, however, was originally envisaged as being of only limited duration. Governments then and indeed for the remainder of the nineteenth century did not believe that it was the state's job to run such institutions. The fact that by the end of the century the number of scientists in government employment had risen considerably did not imply a conscious revision of this point of view so much as the consequence of a series of reactions to new circumstances created by the rise of technology, coupled with increasing expectations among the electorate.[18]

In understanding the relationship between science and the state in Britain during the nineteenth century it is necessary to consider attitudes towards the role of government in general. In the eighteenth century governments could behave towards science in much the same manner as an individual patron but in the nineteenth century the powers of departments and ministers to initiate action were increasingly circumscribed by limits on spending imposed by the Treasury. Its regulations to keep down costs and prevent unauthorized expenditure were, however, a reflection of a more widespread reluctance among much of the community, as well as most administrations, irrespective of political standing, to see public expenditure increase. In trying to stem the tide of the rising cost of government, and fresh burdens on the tax payer, politicians were not solely acting with an eye to vote catching. Their attitude also reflected a deeply-held view on the part of both government and governed, that the duties of central government were largely legislative and prescriptive and should not lead to active involvement except in certain well-defined areas. In his book *Victorian Origins of the British Welfare State* David Roberts[19] shows how, in the decades immediately following the first moves to reform parliament in the 1830s, this ideal began to be gradually eroded in one particular sphere, through concern over welfare which led to public pressure for social reform. Here, owing to the lack of adequate local government machinery, central government bodies were set up to implement new regulations, in spite of the prevailing belief among all shades of political opinion that its powers should not be further extended. Though as Brebner pointed out, 'laissez faire never prevailed in Great Britain or in any other modern state',[20] the idea was nevertheless a powerful one which dominated the thinking of many influential figures in the nineteenth century, and not just with regard to economic affairs.

Attempts to carry out scientific research had to contend with similar attitudes. Nevertheless, in spite of hostility to public spending increases, British government expenditure on science, as in other fields arousing public concern, such as education and health, underwent a gradual increase during the nineteenth century. This happened for a variety of reasons, some philanthropic, others pragmatic, but the fact was that this could happen without undermining basic premises because much of the support was indirect. Here too the need to enforce a growing amount of legislation led the number of scientists directly employed by the government to rise as time went on, rather than any appreciation of the role of science. Two areas important to marine science where this happened were hydrography and maritime meteorology. In the late eighteenth century James Rennell, among others, had called on the government to improve existing surveys of British waters because of shipping losses. This resulted in the setting up of the Hydrographic Department of the Admiralty.[21] In the early stages official surveys were mostly of British waters but gradually they were extended further afield with a view to facilitating the movement of British shipping. Even at this stage many officers involved in this work had wider interests and contributed to a numbers of areas of scientific enquiry, for example the study of terrestrial magnetism.

Two developments in the 1850s obliged government departments to take a more active interest in conditions both at the surface and in the ocean depths. Both elements were to play an important part in the establishment of the science of oceanography in the last quarter of the century, through helping to bring about new methods of investigating the sea and increased public awareness. The discovery in the middle years of the century that telegraph cables could be laid underwater soon led to ambitious schemes to link Europe and North America across the North Atlantic. New techniques were required to obtain accurate charts of the sea bed, so that the best route could be selected for submarine cables. As well as lowering them to great depths, the technology had also so be capable of raising them again for repair.[22] The British government was also anxious to establish telegraphic communication with India. Though cables were laid as commercial ventures, in these early days when the technology was untried and the new companies did not have the resources to carry out surveys for themselves, the government had a double interest in seeing the cables working, on the one hand securing important communication routes and on the other preventing the financial crises that could occur if projects failed. They therefore gave both direct and indirect assistance, in some cases underwriting the cost of cable laying.[23] During the 1850s and '60s naval ships helped to survey cable routes in the Atlantic, Mediterranean and elsewhere and government departments were also involved in improvements in instrumentation.[24] Reliable methods of sounding in oceanic depths and improved methods of measuring temperature and pressure at great depths were necessary for successful cable-laying, but they also formed a prerequisite for the scientific exploration of the deep ocean. The development of these new methods, combined with the use of steam engines, used for propelling the ship and for raising and lowering apparatus, at last made possible the systematic investigation of the depths of the ocean as a routine, though still laborious exercise.

About the same time there was a call for action on an international basis to establish a network for systematic recording of weather and sea conditions. The impetus for this came from Matthew Fontaine Maury of the United States Navy's depot of charts and instruments. In the 1840s he had published wind and current charts derived from data abstracted from ships' logbooks. These were welcomed by shipowners and Maury envisaged a system of observing weather at sea which could be used by the naval and merchant ships of all maritime nations. This plan was discussed at an international congress at Brussels in 1853. In Britain this led to the setting up of the Meteorological Department of the Board of Trade, the forerunner of the present-day Meteorological Office, and maritime meteorology formed a major part of its work in the early years. Marine science however arguably gained as much from Maury's book *The Physical Geography of the Sea*, published at about the same time, which popularized the subject on both sides of the Atlantic.[25]

The support for scientific research entailed in these and similar enterprises did not necessarily mean a change in attitude towards science. In spite of the existence of organizations like the Geological Survey and the Meteorological Office, both carrying out pure science as well as the more practical research for which they had been created, there was still no feeling that it was part of government's role to support scientific research and institutions on a long-term basis. Indeed the prevailing view, even among many scientists, was that such support should come from private sources and that availability of state aid would endanger the purity of intellectual standards. But governments were not monolithic institutions, as Alter and MacLeod have pointed out,[26] and individuals could take a different view. As departments multiplied there were opportunities for politicians and civil servants to interpret policy in a more liberal way in specific circumstances. All, however, were constrained by Treasury control over government expenditure which expanded during the mid-nineteenth century and often operated in a manner hostile to science. The outcome of any one proposal therefore depended on the operation of personalities as well as the nature of the work itself and its possible wider relevance to the community. These circumstances do much to explain why the performance of British marine science was so variable in the late nineteenth and early twentieth centuries.

It may be coincidence that two of the examples given by Roy MacLeod of insensitive Treasury responses to requests for aid for science involved projects in different branches of marine science.[27] However the plight of science in general, as many now saw it, was beginning to cause more widespread unease. People were worried that British education and research facilities were lagging behind her continental competitors, especially Germany, and a royal commission was set up in the early 1870s in consequence.[28] The more widespread discussion of such topics may perhaps have influenced the taking of what still appears an extraordinary decision, to send a naval vessel with a team of civilian scientists on a three-year voyage round the world for purely scientific purposes, yet this is what happened when H.M.S. *Challenger* set sail in 1872.

The more usual response to even modest proposals was wrily outlined only 18 months earlier in an article in *Nature* on the results of a recent Italian expedition:

> In sending a single frigate on such a voyage, the poverty-stricken government of Italy does not hesitate to put on board a band of scientific observers. Does Mr Childers do the same, when he sends his flying squadron round the world composed of the largest and most expensive ships which wealthy England can produce? We are ashamed to say he does not. Any application, even, for a free passage for a naturalist on such an occasion, would receive the stereotyped refusal, and the answer that "my lords" had no funds to devote to such purposes, and no space to spare.[29]

In fact governments might respond favourably, as in the eighteenth century, when projects were put before them for expeditions of a scientific nature, if these were to examine some specific phenomenon and involved the country's international standing, but such help was often given reluctantly, and a special case had to be made out on each occasion. Colonel Alexander Strange wrote to *Nature* in 1870, pointing out sarcastically that the sun was unlikely to defer its eclipse to give the learned societies time to persuade the government to give them funds to go and observe it:[30]

> It may be taken for granted that no encouragement will be afforded by our thrifty rulers to an expedition of sixty-eight astronomers, projected for the quixotic purpose of collecting intelligence not calculated to increase the revenue.

This broadside was part of an ongoing campaign. In the end the government relented and the expedition sailed.[31]

Important contributions to knowledge of ocean fauna, and other aspects of marine science had already been made from naval vessels, either by scientifically-minded commanders such as James Clark Ross or scientists they had invited to sail with them, such as Edward Forbes in the *Beacon*. Young scientists might serve as naval surgeons, as did Thomas Huxley in the *Rattlesnake*. However, such scientific work was usually incidental to the main purpose of the voyage. Biologists were not greatly worried about the lack of opportunity to extend their studies to deeper water because the idea, originally proposed by Forbes, that life could not survive in the sea below a depth of 300 or 400 fathoms, was generally accepted until the 1860s, in spite of some conflicting evidence. This latter encouraged C. Wyville Thomson and W.B. Carpenter to ask the Royal Society to support an appeal to the Admiralty for the loan of a survey ship to dredge at greater depths. The exploration of the marine fauna of British coasts and shallow seas, already in progress for several decades, had been largely the work of dedicated amateurs and yachtsmen who lacked the capability to investigate the deeper water off the continental shelf. This was the origin of the voyages of H.M.S. *Lightning* in 1868 and of H.M.S. *Porcupine* in 1869 and 1870, and these cruises paved the way for the much more ambitious circumnavigation of H.M.S. *Challenger* in 1872-76.[32]

It was due to Carpenter that these proposals did not meet the kind of response outlined above. In the first place he was a man of strong enthusiasms. Once his objective in the earlier voyages, the discovery of life in the depths of the sea, had been achieved it was replaced by an even stronger fascination with the influence of temperature on the circulation of ocean water masses. When his theory was challenged[33] he conceived the idea of a voyage of circumnavigation, comparable to the great exploring expeditions of previous generations, which would extend the new investigations to the major oceans of the world and bring back evidence to support his ideas. A less persistent and well-placed man would have been unable to realize such an ambitious project, but Carpenter not only knew the Cabinet ministers with whom such decisions ultimately rested but also what to say to get them to react favourably. He was also in a position to be able to mobilize scientific opinion effectively.

Writers on the history of government support for science like Peter Alter and J.B. Morrell[34] have seen the *Challenger* expedition as an example of government willingness to support large-scale projects involving national prestige that could not be financed by private enterprise, so long as they did not involve any degree of permanence. The *Challenger* expedition was not unique in this respect and while it was away expeditions were organized to observe the transit of Venus and to attempt to reach the North Pole. Harold Burstyn concluded that it was the opportunity to make deep-sea soundings and other surveys in addition to the scientific work that helped to obtain a favourable decision.[35] This may have provided the practical justification that Victorian ethics needed but Burstyn also stresses the importance of personalities and personal contacts in the success of the enterprise. W.B. Carpenter wrote revealingly to Sir Edward Sabine, President of the Royal Society:

> You are, I feel sure, as anxious as I am myself, that the work which has been so well begun should be prosecuted effectively in the spirit in which the Report of the Dredging Committee was adopted by the Council ... And you are aware that I have already broken ground in the matter with Mr Lowe and Mr Childers, and that any further proposition will be favourably considered by them. My belief is that they are both quite disposed to do anything that is urged upon them by *Scientific Authorities*, if they think that the *Country* will support them. — Now on the latter point I have no question whatsoever. Our Deep-sea researches have excited a very strong interest in the public mind; and every M.P. with whom I have spoken on the subject of them has expressed his willingness to give his hearty support to any plan for the continuance and extension of them, that the Govt may see fit to bring forward. — The question is how best to bring *Scientific Opinion* to bear upon the Ministry; and my experience leads me to feel that personal influence

is often more valuable than official representations it has occurred to me whether this might not be brought to bear very effectively ... Is not our National credit concerned in the prosecution of this enquiry by utilizing the immense resources that our Navy can afford?[36]

Carpenter did not fail to hammer home the desirability of consolidating British primacy in the new field at a time when other scientists in Europe and America were planning their own investigations. Politicians were quite able to withstand the blandishments of scientists; it was public opinion as they conceived it that swayed them. His skilfully orchestrated blend of expectation at home combined with concern for the nation's standing abroad achieved the result he desired.

Carpenter had hoped that a scientific department would be created at the Admiralty to deal with the results. In fact when the expedition returned in 1876 a special organization, the Challenger Office, headed first by Wyville Thomson, scientific leader of the expedition and Carpenter's partner in the earlier voyages, and after his death by John Murray, was set up. The grant was made for five years and there was no suggestion that the office was more than a temporary affair. In the event it was 19 years before the final volume appeared. Murray managed to win several extensions of the grant but in the end he had to complete the publication at his own expense. The fact that the British government had mounted such an expedition did not mean that they accepted any long-term commitment to oceanography. Ironically, European and American scientists were able to reverse the process very successfully in years following, arguing that if the British government had been so generous in its support of the new science their own governments should be no less forthcoming.

There was never any doubt that the *Challenger* expedition was an event of extraordinary significance in the development of oceanography, indeed many see it as the starting point of the modern science. Some aspects of the science were virtually created at that time, in particular the study of sea floor sediments and the study of life in the deep oceans, and though, as we have seen, other aspects of marine science had attracted considerable interest earlier, it was only now that scientists working on different aspects of oceanography began to think of themselves as working within a common discipline. However, though the government honoured its word to Wyville Thomson on the expedition's return and paid for most of the publication costs of its report, the work that it had begun was largely continued in other countries because politicians and civil servants in the

Treasury combined in wanting to keep government costs down and resisting any new forms of expenditure. Thus when Wyville Thomson and John Murray attempted to arrange further small-scale expeditions in British waters after the return of the expedition they were only able to obtain two short voyages, in the *Knight Errant* and *Triton* in 1880 and 1882.[37] Lack of opportunities of this kind was to hamper the development of British oceanography during the next 30 years. Individuals gave generously to science but Britain had no one in oceanography with resources on the scale of Prince Albert of Monaco or Alexander Agassiz in the United States. It was not that the Admiralty, or individual Hydrographers, were themselves hostile in general to work of this kind that prevented more co-operation but the requirement to obtain approval of new items of expenditure with the Treasury.

Scientists were by now seriously worried about the possible effects of this restrictiveness not only on the progress of science itself but on national prosperity in general, but they were also divided about the desirability of the external control that government funding would give, and about the possible effect of a greater availability of funds on the quality of science. Yet if they wished to create scientific institutions and long-term research programmes this is what they had to do. However they were also men of their time and though they might complain about government neglect of science, there were those who shared the general view that science should be independent and finance its operations in other ways. In the marine sciences in the 1870s the way forward was not generally seen as sending out large and expensive expeditions but through the foundation of marine laboratories, or zoological laboratories as they were then more usually called. This followed on from the development of marine biology and its contribution to research in more general areas of biological enquiry, such as embryology, sparked off by Darwin's theory of evolution. Whereas for many years the amateur marine biologist and even the academic had been content to work on his own, ideas both on research and teaching were changing. The introduction of practical work into classes and the expansion of research in the universities meant that laboratory accommodation became essential, and for work on marine life these had to be situated by the sea. Anton Dohrn's Stazione Zoologica established at Naples in 1872 was generally cited as the pattern, though it was not the first of its kind.

British zoologists were anxious to follow the example set by fellow workers on the Continent, and in the U.S.A., and set up marine stations. However, because of the British dislike of state intervention and reliance on private initiative, this proved hard to do. In the early days some hoped that it might be possible to run such places on a small scale with funds raised from donations by the general public. However those institutions that tried this approach were unable to raise the kinds of sums necessary to pay for long-term running costs. The outcome was foreseen by the zoologist E. Ray Lankester who wrote that the British public was more willing to subscribe to a new church than to the needs of science.[38] He wrote:

> They were taught long ago to subscribe to church-building by the example of states and princes. It requires some such initiation to render the subscription lists of zoological stations popular.

Indeed the situation in this country was very different from that in France where Henri de Lacaze-Duthiers, seeking to establish a second station, on the Mediterranean coast to complement his existing foundation in Brittany, could place his plans before the education ministry, confident that help would be forthcoming. He even had rival communes bidding for the installation because they believed it would be beneficial for the neighbourhood to host such an establishment.[39]

Co-operation through existing scientific bodies, or societies specially formed for the purpose, of which the Marine Biological Association of the United Kingdom is the best known,[40] seemed to the best policy but it soon became apparent that organizations relying on public donations alone could not flourish. The problem was that, even though scientists might believe that science should be independent of government, they learned the hard way that private enterprise was unable to provide the funds that were needed, especially for oceanographic work. Individual benefactors made generous gifts to Plymouth and other British marine stations but those which could not attract state aid almost all eventually succumbed.[41] It is no coincidence that Sir John Murray, internationally regarded as one of the leading oceanographers of his day, spent 10 years surveying the Scottish lochs when he would far rather have been at sea. His Scottish Marine Station for Scientific Research failed to win a share of the money being made available for Scottish fishery research and closed for lack of funds. After having to sell his yacht *Medusa*, when paying for the completion of the *Challenger Report* (the government having refused further extension of its grant for this purpose), it was not until 1910 that Murray was able to hire the Norwegian fisheries research vessel *Michael Sars* for a voyage in the North Atlantic, paid for from the fortune he had made from Christmas Island phosphates.[42]

A somewhat more successful combination was achieved as a result of the growing economic importance of fisheries, and the problems pertaining thereto. Here too biologists had expertise to offer and governments were more willing to spend money in obtaining it. Grants for fisheries research began to be made in the 1880s and tided several marine stations over a difficult stage in their development but sooner, in Scotland, or later, in England, this work was concentrated in government institutions. As a result only a few of the independent stations survived. Even the provision of funds for fisheries work was not a foregone conclusion. This was the time when Gladstone was trying to abolish income tax; however, it was also the era when Home Rule for Ireland was being debated and there was a corresponding sensitivity towards Scottish affairs. Fishery questions were causing some anxiety, in particular the effect of trawling. The traditional panacea of royal commissions followed by legislation had been repeatedly tried without assuaging the fishermen's anxieties. In 1882 the Fishery Board for Scotland was reconstituted and began a programme of scientific research.

If dislike of increasing public expenditure was shared by most shades of political opinion, the task of keeping it in check was seen as an almost overriding duty by the Treasury. Roy MacLeod has shown how, working within this framework, unsympathetic individuals could sabotage scientific projects, in the belief that the whole thing was a waste of time. Yet the Treasury could be itself be overridden and this happened with regard to the scientific work undertaken by the Fishery Board for Scotland and by the Marine Biological Association, one of the examples MacLeod uses to illustrate unhelpful Treasury attitudes.[43] The Act setting up the board in 1882 had not given any clear indication as to whether it was supposed to engage in research or not but members chose to assume that it was part of their duties. The Treasury was willing to agree to this, so long as the expenditure remained within the board's own income, derived from the herring brand duty. However, this was required for projects of more direct benefit to fishermen, such as the construction of harbours, and the scientific work soon exceeded its allowance. The first reaction of the Treasury was to instruct the work to halt but shortly after this ruling was reversed and a special grant made for work to continue. Likewise when the Marine Biological Association applied for a grant, the board and the Treasury were obliged to give it a share of the funds.[44]

Parliamentarians were less united than hitherto in their views over questions of this kind. Late nineteenth century Radicals were happy to support state expenditure on deserving projects; Joseph Chamberlain was a good friend of the MBA and may have been responsible for the favourable turn of their affairs.[45] To the Whigs and moderate liberals, this raised 'the spectre of laissez-faire being destroyed and state socialism being constructed before their very eyes'.[46] George Goschen, chancellor under Gladstone and later, after the Liberal party had split over Home Rule, for the Conservatives, hated waste of any kind and sought to keep public expenditure from increasing. In this he did not see eye to eye with Lord Salisbury, the Conservative prime minister, who complained of the effects of Treasury 'parsimony': 'They do not impose their veto at the beginning of a policy, when they might prevent it, but at the tail of the policy, when they can only spoil it.'[47]

Biologists were not just being opportunistic when they sought to get involved in fisheries work. Problems like the disappearance of oysters and the behaviour of fish populations were of scientific as well as economic interest.[48] Indeed not all fisheries scientists were convinced that science was best served in this way. One of them, W.L. Calderwood, wrote:

> I shall venture to express an opinion with regard to the combination of science and fisheries which is justified, I consider, by the experience of past years .... It has ... become the custom to combine the investigation of fishery problems with the study of pure science in the departments of zoology, botany, and physics. A marine laboratory gives facility for both applied and unapplied science; but the results which accrue do not, I think, represent the benefit which might arise if each branch was kept separate from the other. Fishery investigations are directly for the public good, and, therefore according to our modern views, should be, and are, carried on at public expense. With purely scientific research this is not so, and moreover, the publication of purely scientific papers amongst the more commonplace reports of fishery investigations is not to the advantage either of science or fisheries.[49]

This opinion is debatable. It partook of the common misapprehension that applied science can in some way be carried on satisfactorily without more fundamental investigations whereas in fact it can get nowhere without them. Nor was being allied to projects with a practical orientation necessarily harmful to the search for greater understanding of the ocean, as we shall see. What was true in the decades before and after 1900 was that research on fisheries in Britain had obtained an international reputation but her contribution to oceanography since the days of the *Challenger* had been disappointing. Government willingness to fund fishery research and disregard for other aspects of marine science had produced an imbalance which many saw as unhealthy.[50]

Peter Alter argues that it was during the period 1900 to 1914 that signs of a change in attitude towards science on the part of government can be seen, and that it was not, as some have supposed, the crisis of the First World War that made state support of scientific research more acceptable.[51] Developments in marine science support this view though there is evidence elsewhere for the continuance of negative attitudes towards science in general.[52] Perhaps this is because the change was only a partial one. Two important examples occur of state-backed research in marine science with their origins in the period immediately before World War I. Both were linked to specific economic objectives but were also to prove of great importance for the growth of marine science in Britain during the interwar period. For much of the the nineteenth century government funding even for utilitarian reasons was seen as undesirable and needing to be justified by exceptional circumstances. By the early twentieth century, however, the idea had become more acceptable, though still with the proviso that some tangible benefit to the community was expected. The first instance arose out of the establishment in 1910 of a development fund to assist economic growth, particularly in rural areas. This included fisheries and grants were given to several of the independent marine laboratories, such as Plymouth and Millport, enabling them to take on new staff and engage in valuable new research programmes.[53]

The other new development was sparked off by concern over the growth of southern hemisphere whaling in the Falkland Island dependencies. Scientists, in particular Sir Sydney Harmer of the Natural History Museum, were concerned that this fishery would soon be exhausted, like those of the northern hemisphere. They found allies in the Colonial Office who wished to conserve the fishery as a valuable source of revenue to the Falkland Islands and in 1913 an inter-departmental committee on whaling and the protection of whales was set up and a zoologist sent to South Georgia to gain firsthand information. Progress was halted by the outbreak of war but in 1917 a further committee, on research and development in the Falkland Islands dependencies, began its deliberations which eventually led to the setting up in 1924 of the Discovery Investigations. This was the beginning of 14 years of research on whales in the Southern Ocean. The first expedition, in R.R.S. *Discovery*, 1925-27, was combined with shore-based work at whaling stations to collect data on whale biology and a programme of whale marking in the *William Scoresby* to throw light on their migrations. The changing pattern of whaling meant that the original focus on the waters round South Georgia and the neighbouring islands and Antarctic peninsula shifted to cover the whole of the Southern Ocean. In the new R.R.S. *Discovery II* scientists were able to cover the longer distances involved and to study not just the whales and their staple food, Antarctic krill, but also the whole marine environment in which they lived. As with the work financed by the Development Commission, the stated intention was to carry out work for economic purposes, but in the process important basic scientific research was done. When the work of the Discovery Committee was still in its early stages, H.R. Mill, who retained his interest in oceanography (he had worked with John Murray in the 1880s but left because there were no career prospects), commented:

> I sincerely trust that on this occasion, when the organization has been so complete and the Government departments have taken up the work with such enthusiasm, it will be carried on and developed so that once again, as in the days of the *Challenger*, British oceanographers may lead the world. At the present time they have dropped considerably to the rear.[54]

In fact, the work of the Investigations accumulated an unprecedented amount of observations and collections about this area, and enabled British oceanographers once again to make a contribution to new developments in the science, establishing a nucleus of scientific and technical talent in the field which played a major part in re-establishing oceanography as a discipline in Britain after World War II.

It was only by chance that this talent was not dissipated. The Discovery Committee had been established on a temporary basis to throw light on a specific problem; it was never intended to be a permanent institution. Mill feared that expediency might lead to the loss of another generation of marine scientists, as his own and the succeeding one had largely been:

> It is one of the deplorable things in connection with British exploration that great efforts have often been made, but as soon as an efficient staff has been got together with everybody working smoothly ... the whole thing has been dropped.[55]

The work of the Investigations had already been extended far beyond its original intended span because of the changes in whaling. Had it not been for the outbreak of World War II Mill's fears would probably have been realized.

Naval interest in oceanographic research, as opposed to hydrographic surveying, had developed in the early years of the twentieth century as a result of the rise of submarine warfare and the need to find methods of detection. Accurate use of sonar (Asdic) depended on knowledge of the science of underwater accoustics, including the physical properties of the sea water affecting the behaviour of sound waves travelling through it.[56] In addition, during World War I a former marine biologist, G.Herbert Fowler was attached to the Hydrographic Department, where his principal task was compiling charts of salinity and temperature for use by sub-marines. Fowler was one of those whose attempts to participate in deep-sea researches had been frustrated by the impossibility of obtaining adequate support for work at sea.[57] He passionately believed that the kind of work he was doing during the war ought to be continued in the national interest and mounted a campaign for oceanography to be included in the peace-time work of the Hydrographic Department and for more use to be made by naval vessels, and submarines, in collecting data. Attempts to implement his plans largely fell foul of the post-war defence cuts but as a result of Fowler's initiative a marine chemist, D.J. Matthews, was attached to the department during the inter-war period.

During World War II the navy expanded its research effort, seeking to improve methods of mine and submarine detection and making use of new apparatus developed overseas, in particular the bathythermograph. The need to forecast sea conditions for landings in North Africa focussed attention on waves and swell, phenomena familiar enough to sailors but poorly understood.[58] In 1944 a team of scientists (Group W), led by George Deacon, who had formerly worked for the Discovery Committee, was set up at the Admiralty Research Laboratory to undertake fundamental research on sea waves and swell.[59] While the war was still in progress pressure began to be applied by various interested parties to ensure that when the war was over proper provision should be made for the continuation of oceanographic research along the lines being pursued in other countries. Hydrographers, fisheries scientists and university representatives individually and through the Royal Society pressed for a national oceanographic institute to be established on the grounds that fundamental research into the ocean sciences was in the national interest, both on military and economic grounds. In spite of winning the support of the relevant government departments it was not until 1949 that the National Institute of Oceanography (now the Institute of Oceanographic Sciences) was set up, incorporating marine scientists from Discovery Investigations, Group W and the Hydrographic Department.

The subsequent fortunes of oceanography are outside the scope of this paper. The modern science involves state support at all levels and indeed goes further, with major research projects being undertaken on a basis of international co-operation. If one accepts what Harold Burstyn says, that 'oceanography has always been either "big science" or no science at all',[60] this was an inevitable process. This is not to say that scientists today do not face problems, though these may be of a different variety.[61] If marine scientists had had greater freedom of action in earlier days oceanography might have developed differently. However, the concentration on economic objectives did encourage important theoretical work in fields which might not have otherwise attracted attention. To say this is not necessarily an argument in favour of such constraints. Obviously practical constraints must apply to all fields of human activity and in science money is no substitute for ideas, but the unnecessarily harsh restrictions imposed on British scientists in the nineteenth century, through what amounted almost to an obsession with an ideal of non-intervention which in practice had never existed, led, with the one glorious exception of the *Challenger* expedition, to a waste of talent and enthusiasm. During the period principally covered by this paper it did prove possible in some countries, notably Sweden, to finance marine research on a private basis but in Britain this was never very successful. Nineteenth-century commentators did not blame governments alone for this, believing that British indifference if not hostility to science lay behind such attitudes. No one expected that science should be supported purely for knowledge for its own sake; what they hoped for was some appreciation and recognition that understanding of fundamental processes is necessary in a world relying increasingly on science and technology. In the late nineteenth century there was not the same degree of urgency in learning about the marine environment as is being experienced in the late twentieth century but in a nation which owed so much of its wealth and prosperity to the sea contemporaries found it surprising that a small proportion of this could not be devoted to obtain a better knowledge of this element.[62]

## NOTES

1    Margaret Deacon, *Scientists and the Sea, 1650-1900: a Study of Marine Science* (London, Academic Press, 1971). Much has been added to the history of oceanography during the last 20 years, but most of the contributions have been on the nineteenth and twentieth centuries and earlier interest in the science of the sea has received little attention.

2    Harold L. Burstyn, 'Pioneering in large-scale scientific organisation: the *Challenger* expedition and its report. I. Launching the expedition,' *Proceedings of the Royal Society of Edinburgh*, B LXXII (1972), 47-61, on p. 48. On p. 47 he refers to it as 'probably the major single research project of all time'.

3    Peter Alter, *The Reluctant Patron. Science and the State in Britain, 1850-1920*, translated by Angela Davies (Oxford, Hamburg, New York, Berg, 1987).

4    William Oughtred, *The Circles of Proportion* (London, 1633), part 2, 55-6, quoted in Deacon, *Scientists and the Sea*, 70.

5    Deacon, *Scientists and the Sea*, 75, 85.

6    Louis Ferdinand Marsigli, *Histoire Physique de la Mer* (Amsterdam, 1725).

7    Marsigli, *Histoire Physique*, 47. Marsigli realized that government aid would be needed though he puts this in eighteenth-century terms. The phrase he uses, 'quelque Prince, amateur & protecteur des Sciences' in fact exactly describes Prince Albert I of Monaco who, by coincidence, was a leading oceanographer and patron of science nearby in the late nineteenth and early twentieth centuries.

8    Deacon, *Scientists and the Sea*, 99.

9    Harry Woolf, *The Transits of Venus: a Study of Eighteenth Century Science* (Princeton, University Press, 1959), 81; J.C. Beaglehole, *The Life of Captain James Cook* (London, Hakluyt Society, 1974) extra series, XXXVII, 101-02. Beaglehole is concerned with the campaign by scientists to ensure observations of the 1769 transit, the reason behind Cook's first voyage in H.M.S. *Endeavour*. Hornsby recommended that a settlement should be made in the Pacific for commercial reasons. Woolf demonstrates how the British government of the time responded to knowledge of what other countries were planning and was sensitive to national prestige in this area.

10   Themes of this kind are investigated in, for example: Alan Frost, 'Science for political purposes: European exploration and the Pacific Ocean, 1764-1806,' in Roy MacLeod and Philip F. Rehbock, eds, *Nature in its Greatest Extent: Western Science in the Pacific* (Honolulu, University of Hawaii Press, 1988), 27-44.

11　See Norman J. Thrower, ed., *The Three Voyages of Edmond Halley in the Paramore, 1698-1701* (London, Hakluyt Society, 1981), series 2, CLVI.

12　David Mackay, *In the Wake of Cook: Exploration, Science and Empire, 1780-1801* (London, Croom Helm, 1985).

13　Ann Savours, ' "An interesting point in geography": the 1773 Phipps expedition towards the North Pole,' *Arctic*, XXXVII (1984), 402-28.

14　A.L. Rice, 'The oceanography of John Ross's Arctic expedition of 1818; a reappraisal,' *Journal of the Society for the Bibliography of Natural History*, VII (1975), 291-319.

15　Deacon, *Scientists and the Sea*, 220-3, 238-40.

16　*Ibid.*, 280-1.

17　Philip F. Rehbock, 'The early dredgers: "naturalizing " in British seas, 1830-1850,' *Journal of the History of Biology*, XII (1979), 293-368; A.L. Rice and J.B. Wilson, 'The British Association Dredging Committee: a brief history,' in Mary Sears and Dan Merriman, eds, *Oceanography: the Past* (New York, Springer, 1980), 373-85. For a general account of the founding and early years of the association see, Jack B. Morrell and Arnold B. Thackray, *Gentlemen of Science: Early Years of the British Association for the Advancement of Science* (Oxford, Clarendon Press, 1981).

18　Roy M. MacLeod, 'The Royal Society and the government grant, 1849-1914,' *Historical Journal*, XIV (1971), 323-58. For the early history of the Geological Survey, see: John Smith Flett, *The First Hundred Years of the Geological Survey* (London, 1937).

19　David Roberts, *Victorian Origins of the British Welfare State* (New Haven, Yale University Press, 1960).

20　J. Bartlett Brebner, 'Laissez-faire and state intervention in 19th-century Britain,'*Journal of Economic History*, VIII (1948), 59-73, 60.

21　Clements R. Markham, *Major James Rennell and the Rise of Modern Geography* (London, 1895). Rennell's work on ocean currents is described on pp. 146-70. He was invited to become the first Hydrographer of the Navy but declined because it would have meant giving up his other geographical work, on the mapping of Africa, ancient geography etc.

22　For the most comprehensive account of how this was done and the other instrumentation involved was developed see, Anita McConnell, *No Sea too Deep: the History of Oceanographic Instruments* (Bristol, Adam Hilger, 1982). See especially chapter 5, pp. 49-72 on 'The grand Victorian technology: submarine telegraphy.'

23　R. Murdoch Smith, 'Sketch of the history of telegraphic communication between the United Kingdom and India,' *Scottish Geographical Magazine*, V (1889), 1-11.

24 Later the cable companies became sufficiently large and wealthy
to have ships and carry out surveys of their own (and incidentally
collect a large amount of data of potential interest to scientists, as
their successors the oil exploration companies do today.) See,
Anita McConnell, 'The art of submarine-cable laying: its contribution to physical oceanography,' in Walter Lenz and Margaret
Deacon, eds, *Ocean Sciences: Their History and Relation to Man* (Proceedings of the 4th International Congress on the History of
Oceanography, Hamburg, 1987), *Deutsche Hydrographische
Zeitschrift*, Ergänzungsheft, series B, XXII (1990), 467-73.

25 Jim Burton, 'Robert FitzRoy and the early history of the
Meteorological Office,' *British Journal for the History of Science*, XIX
(1986), 147-76.

26 Alter, *The Reluctant Patron*, 62; Roy M. MacLeod, 'Science and the
Treasury: principles, personalities and policies, 1870-85,' in G.
L'E. Turner, ed., *The Patronage of Science in the Nineteenth Century*
(Leyden, Nordhoff, 1976), 115-72.

27 MacLeod, 'Science and the Treasury', 146, 148. The request to
fund tidal observations was turned down in spite of its 'obvious
utilitarian interest'. Grants for fisheries research were unwillin-
gly approved because of outside intervention. Alter, *The Reluctant Patron*, 62, states, 'From at least the middle of the century it is
clear that the state was increasingly ready to offer financial aid to
specific areas of science, and to exert patronage over the most
diverse scientific activities, primarily for utilitarian reasons'. This
is true but, as MacLeod shows, such help was by no means
guaranteed.

28 The Royal Commission on Scientific Instruction and the Advancement of Science (often referred to as the Devonshire Commission because it was chaired by the Duke of Devonshire) began
hearing evidence in 1872 and published its report in 1875. It concluded that in principle the state should support science because
whatever may be the disposition of individuals to conduct researches at their own cost, the advancement of modern science
requires investigations and observations extending over areas so
large and periods so long that the means and lives of nations are
alone commensurate with them. (Eighth Report of the Royal
Commission on Scientific Instruction ... [c. 1298] 1875, xxviii, 417,
24).

29 *Nature*, III (1871), 268-9. The quotation comes from a review of
E.H. Giglioli's 'Noto intorno alla distribuzione della Fauna
invertebrata nell'oceano, presse durante un viaggio intorno al
Globo, 1865-68'. This is signed P.L.S., probably the zoologist
Philip Lutley Sclater.

30    Alexander Strange, 'The government and the eclipse expedition,' *Nature*, II (1870), 512-13.

31    *Nature*, III (1870), 92.

32    Deacon, *Scientists and the Sea*, 306-16, 333-5.

33    *Ibid.*, 318-28.

34    Alter, *The Reluctant Patron*, 66. J.B. Morrell, 'The patronage of mid-Victorian science in the University of Edinburgh,' in G.L'E. Turner, ed., *The Patronage of Science in the Nineteenth Century* (Leyden, Nordhoff, 1976), 53-93, on p. 83.

35    Harold L. Burstyn, 'Pioneering in large-scale scientific organisation', 50. He observes (p. 53) that the 'Treasury's usual acerbity towards new expenditure was surprisingly muted in the case of the expedition.' Burstyn quotes George Goschen, then First Lord of the Admiralty, reminding Robert Lowe, Chancellor of the Exchequer, that a favourable reply had been agreed by the Cabinet. One wonders whether Goschen would have been so ready to do this when he himself became chancellor. One thing that may have influenced both men is that they were elected Fellows of the Royal Society about this time.

36    Royal Society, Sabine Papers. W.B.Carpenter to Edward Sabine, 3 November 1869.

37    Margaret Deacon, 'Staff-Commander Tizard's journal and the voyages of H.M. Ships *Knight Errant* and *Triton* to the Wyville Thomson Ridge in 1880 and 1882,' in Martin Angel, ed., *A Voyage of Discovery* (Papers presented to Sir George Deacon on his 70th birthday), (Oxford, Pergamon, 1977), 1-14.

38    E. Ray Lankester, 'An American sea-side laboratory,' *Nature*, XXI (1880), 477-9. This was true in other fields too: Alter, *The Reluctant Patron*, 60, writes, 'Ultimately private patronage could not supply enough money to finance large research undertakings and institutes in the long term'.

39    H. Lacaze-Duthiers, 'Les progrès de la Station Zoologique de Roscoff et la création du laboratoire Arago à Banyuls-sur-Mer,' *Archives de Zoologie Expérimentale*, IX (1881), 542-98.

40    A.J. Southward and E.K. Roberts, 'One hundred years of marine research at Plymouth,' *Journal of the Marine Biological Association of the United Kingdom*, LXVII (1987), 465-506.

41    Margaret Deacon, 'Crisis and compromise: the foundation of marine stations in Britain during the late 19th century,' *Earth Sciences History* (in press).

42    Harold L. Burstyn, 'Science pays off: Sir John Murray and the Christmas Island phosphate industry, 1886-1914,' *Social Studies of Science*, V (1975), 5-34. This article investigates Murray's claim that by 1913 the British government had recouped its expenditure

on the *Challenger* expedition through taxes paid by his company. Burstyn concludes that the claim was exaggerated, but points out that the amount paid would have been much greater if the company had not been prevented from starting work earlier while conflicting claims were settled.

43　MacLeod, *Science and the Treasury*, 148.

44　Margaret Deacon, 'State support for "useful science": the scientific investigations of the Fishery Board for Scotland, 1883-1899,' in Harry N. Scheiber, ed., *Ocean Resources: Industries and Rivalries since 1800* (Working papers for the 10th International Economic History Congress), (Berkeley, University of California Center for the Study of Law and Society, 1990), 15-29.

45　Southward and Roberts, 'One hundred years of marine research', 160. See also, MacLeod, 'Science and the Treasury', 156, for the change of Treasury attitude which made them willing to allow the MBA a capital grant of £5000 towards the building of its laboratory.

46　Thomas J. Spinner, *George Joachim Goschen: the Transformation of a Victorian Liberal* (Cambridge, University Press, 1973), 92.

47　Quoted by Spinner, *Goschen*, 164-5.

48　E. Ray Lankester, 'The value of a marine laboratory to the development and regulation of our sea fisheries,' *Journal of the Society of Arts*, XXXIII (1885), 749-55.

49　W.L. Calderwood, 'British sea fisheries and fishing areas, in view of recent national advance,' *Scottish Geographical Magazine*, X (1894), 69-81.

50　C.A. Kofoid, 'The biological stations of Europe,' *United States Bureau of Education Bulletin*, no. CCCCXXXX (1910) 4, 144, says, 'The scientific fisheries work done by the British stations is unsurpassed in its excellence ... but the strictly scientific phases of the stations' activities too often suffer for lack of adequate financial support and from consequent loss of scientific interest'.

51　Alter, *The Reluctant Patron*, 6, 132.

52　See Russell Moseley, 'Government science and the Royal Society: the control of the National Physical Laboratory in the inter-war years,' *Notes and Records of the Royal Society*, XXXV (1980), 167-93, especially p. 173 where he says that the Treasury attitude 'remained substantially that of pre-war days: research must ultimately become financially self-supporting'.

53　Eric L. Mills, *Biological Oceanography: an Early History, 1870-1960* (Ithaca and London, Cornell University Press, 1989), 206-07. As a consequence of this, he says, Plymouth became 'the center of research on the plankton cycle for nearly twenty years.' Mills is concerned to examine the reasons why work on productivity in the sea should have declined in its original home at Kiel in Germany and the initiative passed to the U.K. This was due to other factors than those under discussion in this paper. As Roy MacLeod points out in his introduction to 'The Royal Society and the government grant,' p.323, scientific creativity does not work to order and the history of science is therefore not the same thing as the history of scientific institutions. However external conditions 'may affect the rate and direction in which [scientific] discovery may occur'. The Marine Biological Association would have been unable to engage in this work without the opportunity first to expand its scientific staff.

54 These remarks by Hugh Robert Mill were made during discussion following a lecture at the Royal Geographical Society in April 1928; see, A.C. Hardy, 'The work of the Royal Research Ship *Discovery* in the Dependencies of the Falkland Islands, *Geographical Journal*, LXXII (1928), 209-34, on p. 231.

55 *Ibid.*

56 Willem Hackmann, *Seek and Strike. Sonar, Anti-Submarine Warfare and the Royal Navy, 1914-1954* (London, H.M.S.O., 1984).

57 Margaret Deacon, 'G. Herbert Fowler (1961-1940): the forgotten oceanographer,' *Notes and Records of the Royal Society*, XXXVIII (1984), 261-96.

58 Henry Charnock, 'Sea and swell forecasting for operational planning,' in Brian D. Giles, ed., *Meteorology and World War II* (Birmingham, Royal Meteorological Society, 1987), 22-30.

59 Henry Charnock, 'George Edward Raven Deacon, 1906-1984,' *Biographical Memoirs of Fellows of the Royal Society*, XXXI (1985), 113-42.

60 Harold L. Burstyn, 'Science and government in the nineteenth century: the *Challenger* expedition and its Report,' *Bulletin de l'Institut Océanographique*, special no. II (1968), 2, 603-11, on p. 604.

61 Heinz Riesenhuber's 'Perspectives of modern research policy in Europe,' *Science and Public Affairs*, V (1990), 31-43, contains a present-day politician's warning that the current trend of bureaucratic direction of science may prove counter-productive.

62 For example, Lyon Playfair, during discussion of Lankester's paper, 'The value of a marine laboratory', 757, remarked that, 'It was really disgraceful that a country washed all round by the sea - a maritime nation *par excellence* - should be the only one not conducting these scientific researches'.

# MODERN U.S. PACIFIC OCEANOGRAPHY AND THE LEGACY OF BRITISH AND NORTHERN EUROPEAN SCIENCE

## Harry N. Scheiber

In the decade immediately following World War II, American fisheries scientists and oceanographers inaugurated a cluster of large-scale deepwater research projects in the Pacific Ocean that would quickly make an extraordinary mark in the history of modern ocean science.[1] These projects not only mobilized government funding, ships, gear, and scientific talent on a scale never before known in American oceanography, so as to produce a vast amount of new data on all aspects - physical, chemical, biological, and climatological - of the Eastern and Central Pacific regions and their living resources. But they also went far toward effecting a basic transformation in the ways in which modern ocean scientists conceptualize and investigate marine environments and the species that inhabit them. The postwar Pacific projects embodied, in a word, the ecosystem approach to environmental studies that since the 1880s had been an oft-expressed ideal in the scholarly literature and theory of 'ecology' as a distinctive field of science - but an ideal, or model, only rarely carried fully into practice in deepwater fisheries research.[2]

The projects in question included, first, 'CalCOFI,' a cooperative investigation of the decline in California sardine harvests, involving federal, State of California, and academic scientists in what became a massive and comprehensive study of the marine ecosystem of the California Current.[3] Second was the Pacific Ocean Fisheries Investigation (POFI), a project of the federal government that was based in Hawaii and dedicated to exploration for tuna in the Central and Eastern Pacific; this project, too, quickly expanded to cover a vast area of the Pacific region, and it was responsible both for opening a rich new tuna fishery and coincidentally for important discoveries in physical oceanography.[4] Third was the Inter-American Tropical Tuna Commission research program, inaugurated under terms of conventions with Mexico and Costa Rica, later expanded to include other nations, which engaged in basic biological and ecological research on the tuna waters off the Continental Shelf of Central and Latin America; in the 1960s, the research became the basis for a management program for tuna in that region.[5]

Taken together, these projects received cumulative funding in the tens of millions of dollars and involved the use of some ten ships engaged fulltime in oceanographic and fisheries research. Their studies were coordinated at sea and in laboratories, moreover, with those being conducted by the West Coast state agencies and by fisheries scientists in the region's universities and academies. The overall effort represented at least a five-fold increase in the number of ships and ten-fold the appropriation allocated by the United States to deep-sea Pacific marine research in the immediate prewar period and in 1945-47.[6]

The present study explores two facets of the history of these postwar projects that are of particular interest to students of maritime history. Part One will examine the relationship between the rich heritage of ecosystem study - the legacy of ecological research, developed in Northern European and British science beginning in the late nineteenth century - and the direction that the American ocean scientists gave to their Pacific Ocean projects after the Second World War.[7] An intriguing feature of the Pacific projects was their immediate acceptance of - and attempt to apply in a comprehensive way - the European ecosystem research model. Was this, the historian must ask, an intellectual 'rediscovery' of sorts - the revivification of a scientific approach long neglected, during years of regional parochialism in the western United States, and now suddenly perceived as a plausible methodology for the new Pacific projects? Or were there more direct influences and lines of continuity that connected, in palpable ways, these American initiatives after 1945 with the Anglo-Scottish and Northern European scientific traditions in marine studies?

In Part Two, the analysis identifies some remarkable parallels between the pattern of historical forces that converged in the late 1940s to produce the new U.S. Pacific research programs, and the pattern that in an earlier era had produced the advances in organization of marine science in Scotland and England, such as Margaret Deacon writes about in her contribution to this volume. Throughout the history of ocean science in Europe and America alike, it has only been at a few critical moments that major new initiatives in marine research have succeeded in mobilizing the funding and momentum necessary to advance the field significantly.[8] Hence the analysis of these rare moments of spectacular success can perhaps cast light on major issues confronting us now, if we are concerned about maintaining the vitality of oceans research.[9]

## I. Ecology, Fisheries Oceanography, and the Course of Scientific Influence

The ecosystem research model in the field of ecology was first developed mainly by marine biologists in the late nineteenth century. True to that tradition, a few of Northern Europe's and Great Britain's greatest marine scientists, led and exemplified by Johan Hjort of Norway and later by Michael Graham of Britain, did seek to develop fisheries science based on systematic investigation of the marine environment in relation to fishery species and populations within discrete spatial areas of the Atlantic Ocean and North Sea.[10] Such research was, of course, exceedingly difficult because of technological limitations of gear and instrumentation; and because of the intractable problems associated with deepwater research across large distances involving study of fishery species whose size, spatial distribution, and migration patterns - and, as was later discovered, whose subdivision into populations within species - were largely unknown and exceedingly difficult to establish.[11]

Meanwhile, the immediate problems associated with fisheries in danger of depletion set up pressures for a much more practical approach, with some hope of immediate pay-offs, to fisheries research and management. The fishing industry sometimes demanded, but more typically accepted most reluctantly, research programs that looked toward assessment of fishery stocks for purposes of introducing management regimes designed to conserve them at sustainable levels. This need for applied research resulted in a widening gap between biological oceanography - which did remain concerned with the ecosystem approach of ecology - as a field of 'pure' science, and fisheries management research, which moved more and more deeply into exclusive concern with harvest productivity analysis - i.e. the analysis of inputs (number of ships and labourers, time on the water, gear employed, etc.) in relation to output (the volume of the harvest), a subject to which we return shortly. This divergence of lines of research characterized even fisheries management studies in the Northern European nations and Great Britain, where early ecological science had been given the most serious attention and had made the greatest progress from the 1880s to the First World War.[12] Thus Dr. Robert McIntosh, an historian of the science of ecology, has speculated that although biological oceanography in the early years was clearly pursued on the model of ecological science, and was central to research efforts in ecology, 'the link with commercial fisheries and the attendant institutionalization of marine studies in government-supported organizations' produced the divergence on which we have remarked here, largely 'separat[ing] fisheries studies from the nascent science of ecology.'[13]

During the interwar period of the 1920s and 1930s, the divergence of marine fisheries management studies and ecological marine science in the United States had become virtually complete.[14] Among projects of significant scope, only the studies of the Georges Bank region of the Atlantic, conducted on the *Atlantis* during 1939-41 by the Woods Hole Oceanographic Institution, even approached the ideal of a holistic environmental study of a spatial ocean region and its environment. Even the Georges Bank investigations, however, were small-scale; after an initial period of interest in the population dynamics of haddock they became almost exclusively devoted to the plankton and its relationship to the energy chain, rather than to systematic analysis of the relationship of physical and chemical phenomena to fisheries populations and their dynamics.[15]

Marine fisheries studies on the Pacific Coast of the United States from the early 1920s to World War II had by no means languished; but scientific efforts in Pacific Coast institutions exemplified the divergence, in ocean science, of fisheries research away from the ecosystem work that was being more and more concentrated on ocean chemistry in relation to the plankton. Pacific Coast fisheries management research took a narrow focus, moving progressively toward greater emphasis upon harvest productivity analysis.[16]

The marine laboratory of the State of California, for example, had developed a reputation in the 1920s for brilliant work on marine fisheries; but, with few exceptions, the main achievements of the laboratory in the interwar years was in the development of productivity (or yield) analysis as a tool for management. Such research involved two main objectives: to establish the mathematical relationship between fishing effort (i.e. inputs) and the volume of harvest; and to analyse - through use of fish-tagging and then careful analysis of the specimens harvested in the commercial catch - the age (year-class) and size distribution and other characteristics of the aggregation of fish taken. These approaches offered a statistical basis for establishing quotas on the catch or other forms of limitation of fishing effort (gear-design regulations or prohibitions, seasonal restrictions, number of days on the water per vessel, etc.). Only marginally did the fisheries scientists in the California and other West Coast state management agencies concern themselves with empirical data collection or theory relating to the ecosystem in which the fish were being taken, or the interrelationship of the fishery populations to that environment. Even such obvious issues as interspecies competition for food supply presented problems beyond the scope of the research these scientists' resources would support.[17] Physical oceanography was largely outside the realm of study; indeed, during the interwar years, 'fisheries oceanography,' the phrase employed in this paper's title, would have

been entirely inappropriate as a description of what fisheries management scientists were doing. Only in later years, as the result of the
transformation of fisheries research methodology that was effected in
the U.S. Pacific projects we are discussing, did the phrase become
accurately descriptive.[18]

Dr. William F. Thompson, who played the key leadership role in
developing the California state marine laboratory and its research
program, was also responsible for two other highly important fisheries
management initiatives in the 1920s and 1930s. First was the research
program of the international halibut commission, operated under a
Canadian-U.S. treaty, which - on the basis of Thompson's investigations of the causes of halibut depletion - instituted a halibut fishing
regime that was widely regarded as being responsible for engineering
the recovery to commercially viable levels of a fishery that had nearly
died out from uncontrolled over-harvesting on a virtually slash-and-
burn basis.[19] A second program led by Thompson was the research of
the International Pacific Salmon Commission, also formed by treaty
between the United States and Canada, whose work carried over into
(and overlapped with) research by the Fisheries School of the
University of Washington at Seattle, of which Thompson was director.
After 1947, studies were also conducted by the Salmon Research
Institute, an industry-sponsored independent agency which also was
headed by Thompson.[20]

The spectacular success that his yield-analysis approach had produced in responding to the halibut depletion problem - a management
regime that saved the industry in British Columbia, Washington, and
Oregon - gave Thompson enormous influence with industry and
policy officials as well as scientists on the West Coast. He was interested in working on fisheries under commercial exploitation; the tagging efforts and analysis they supported were part of a highly practical larger enterprise of getting the industry to cooperate in establishment of quotas and regulations based on his year-to-year harvest
analysis.[21] To achieve the immediate results that were demanded of
him, and with the small quota of days at sea and the technology available to him, Thompson could scarcely afford to think in terms of
trying to achieve a fuller analytical picture of the ecosystem in which
his halibut, salmon, and (in California) sardine and other species
dwelt. To be sure, he and his collaborators did engage in some studies
of larvae and juveniles in the salmon fishery, seeking clues as to
reproduction rates and survival mechanisms; and he also expended
some time on studies of the chemistry and temperature of the fishery
waters.[22] Thompson's consuming interest, however, was always to
gather data and produce reliable calculations that would maximize the
chances that the scientists would come up with the 'right' quotas in
setting the year-to-year rules of the regime.[23]

In attempting to explain why the various elements of oceanographic science diverged as they did in the directions and methodologies of research, Eric Mills has argued, it is not enough to say (as McIntosh did) that fisheries management scientists were controlled by the pressures of industry and imperatives of politics. 'There is a grain of truth in this generalization, but the emphasis is wrong,' Mills writes. 'We must look instead - or at least equally - to the kinds of men and disciplines that were brought to bear on the quantitative biology of the seas between the late 1870s and the 1950s.'[24] Does the case of William F. Thompson offer support for Mills' contention about why fisheries studies branched off in a divergent direction from ecological research?

Thompson was, after all, not only responsible for nearly all the major fisheries-research initiatives on the West Coast in the interwar period; he also was the doctoral dissertation adviser of half or more of the principal scientists who carried on his work in the region after the war.[25] Because of the long shadow that Thompson, as a result, cast over fisheries research and management in the American agencies on the Pacific Coast, his career and its impact may seem at first glance to constitute impressive evidence in favor of Mills' argument stressing individuals and their influence. Thompson gave first priority to harvest analysis, and he taught its central importance to his students; he achieved a stunning success with it that made the halibut program the model for American agencies' management of Pacific stocks for several decades; and he enjoyed the enthusiastic support of industry and policy officials for his approach.[26]

Characteristically, then, when in 1946-47 one of his most distinguished students, the biologist Wilbert McLeod Chapman, was promoting a new program of research on Pacific tuna - a species about which little was known and which then was not fished commercially throughout the largest part of its presumed habitats in the Pacific Rim as a whole - Thompson very sternly warned him that significant research results could only be achieved when there was a commercial fleet in operation and cooperating with the scientists in a program of harvest analysis. It was almost unthinkable to him, one could read from Thompson's arguments, that scarce resources in ships, instruments, and scientists should be expended on so problematic and expensive a venture as studying tuna stocks that were swimming in a free natural condition in an extensive ocean habitat.[27] Doubtless Thompson would have reacted even more negatively had Chapman revealed that he had in mind a study of the ecosystem, and not merely an intensive exploratory hunt for evidence of 'the abundance of species in different parts of their ranges.'[28]

Meanwhile, it should be noted, the marine biologists in the academic institutions of the West Coast - that is, the University of

*Plate 1.*    Portrait of Roger Randall Dougan Revelle in laboratory, c.1958. (Courtesy of Scripps Institute of Oceanography Archives)

California at Berkeley, Scripps Institute of Oceanography (SIO), Stanford University, and even the Friday Harbor biological laboratory of the University of Washington - had eschewed nearly altogether in the 1930s the kind of fisheries yield research the scientists employed by the management agencies were pursuing in the Thompson style. Although Scripps Institute of Oceanography (SIO), actually stressed marine biology until 1937, and in an earlier era (under leadership of William Ritter) had shaped its biological program on ecological lines, the SIO studies had little to do with fishery dynamics. As one scholar at SIO recalled, the biologists and oceanographers at that institution regarded 'fisheries studies' as something very distinct from 'marine biology' as a basic science: 'There was really no thought of fisheries [at SIO] in the 1930s, though quite a few of the faculty were seriously concerned with marine biology' - and although one of the faculty members, Richard Fleming, later became an important figure 'in developing the holistic view of the ocean.'[29]

The isolation from fisheries science was true to an even greater degree, if possible, with regard to the physical oceanographers. As one fisheries biologist declared privately in 1947, 'The trouble is that most good physical oceanographers not only have no conception of the importance of fisheries problems, but they are generally of the belief that they will soil their hands by touching anything except "pure" science'.[30] Looking at things from the other side of the fence, the eminent geophysical oceanographer Dr. Roger R.R. Revelle (see Plate 1) recalled that Thompson and his followers:

> had no touch with oceanography. What they thought important was species life history - abundance - and the regulation of fisheries production. What [the Pacific project leaders] thought about was that the ocean is an ecosystem ... We thought of marine biology as a much more basic enterprise - dealing with the whole living wealth of the sea and its relationship to marine environment.[31]

As for the physical oceanographers generally - not only the group at Scripps Institution - Revelle recalled that they shared the common view that for any oceans research, 'You've got to get the physics right first'; in any event, they were 'not interested much in biology', let alone fisheries management issues.[32]

This brings us back to the intriguing matter of influences upon the postwar West Coast fisheries scientists other than the predominant one associated with Thompson and productivity or yield analysis. For the new projects inaugurated in the Pacific all represented a very significant departure from the prevailing mode of Thompson-style research.

43

The first in time of these projects, CalCOFI, was initiated in 1947 as a cooperative enterprise involving (as noted earlier) the U.S. Government fishery agency, the State of California's marine laboratory, and the University of California's SIO facility; additional small contributions were made by Stanford University and the California Academy of Sciences in San Francisco. Its purpose was to investigate the causes of the dramatic decline in abundance of the sardine in the California Current.[33] Even in the sardine project's initial months, however, the project's design was expressed by the scientists in charge in terms that can only be described as those of ecosystem research: they articulated a vision that embodied the putative transformation of the scope and content of scientific method in ocean science, anticipating by more than a decade the new push into research on 'environments' that would give the science of ecology a central and unifying role in natural-resources studies in America.[34]

Thus, far from hewing to the Thompson formula that dictated a strictly delimited and instrumentalist (management-oriented) research, with yield analysis at its core, the CalCOFI scientists spoke from the very start about the need 'to establish the relationship between oceanographic fluctuations and the concomitant fishery phenomena [requiring] a continuous record of conditions in both ... physical oceanography and fisheries.'[35] Revelle of SIO, the lead institution in the consortium of agencies and principal organizer of the physical oceanographic and meteorological work in the sardine project, thus contended in 1947 that a new conceptual framework of biological studies in relation to the ecosystem would be needed for CalCOFI - what Revelle termed would be a project to conduct dynamic analyses of the processes in the sea, that is, the cause and effect relationships which affected sardine production.

> In the past, oceanographic research has been concerned primarily with the description of *average* conditions prevailing in the sea. The investigation upon which we are about to embark poses a new and more difficult problem, that is, of studying the nature and causes of *variations* from the average conditions. The present is a good time to start such an investigation, because obviously [on account of the declining harvests] we are in a period of major departure from the average conditions, at least insofar as the distribution the sardine population is concerned ....
>
> In attacking a problem of such magnitude all possible scientific tools and methods will have to be employed. It will be necessary first to describe as completely as possible

the existing oceanographic and biological situations; second, to establish empirical statistical correlations between the various environmental and biological factors; and third, and most important, to make dynamic analyses where possible of the processes in the sea, that is, the cause and effect relationships which affect sardine production. Wherever such a dynamical analysis of a particular aspect of the problem can be made, a great saving in time required for a solution will be effected over the 'brute force' method of statistical correlation which requires a long series of observations for validity.[36]

The vision that Revelle and his colleagues were setting forth thus was one that called for interdisciplinary research in a holistic mode: its focus was to be the marine ecosystem.[37] As Revelle argued, it had already been learned from experience in research in geology and geophysics that 'far more productive results are obtained by complete analysis of all the factors which exist in a particular situation than by a statistical treatment of a few factors in many situations'. He thus proposed that the dynamics of the sardine in its California Current environment, across a vast sweep of space in the Pacific Ocean offshore of Mexican and U.S. California, should be studied in precisely the same way.[38] And in fact, this is what the CalCOFI project, blessed with new and well-outfitted ships and unprecedented levels of funding, proceeded to do - on a scale so great that the California Current became within a few years the most intensively studied deepwater ocean region in the world.[39] Revelle's boast that SIO and its collaborators could 'make the entire Pacific our oyster' was not, in the end, misplaced.[40]

In this respect, the CalCOFI sardine project represented a radical departure from the prevailing mode of fisheries management research not only on the American Pacific Coast but even, by then, in Great Britain and the European continent. It was, in effect, a return to the model of an old and noble tradition of fisheries work that had been exemplified at the turn of the century by Hjort and the early leadership of the International Council for the Exploration of the Seas (ICES)- work carried on by Michael Graham and others in the UK - which had sought to study ecosystems and to integrate fisheries biology and management concerned with broadly conceived, interdisciplinary research on the entire web of environmental relationships.[41]

In similar fashion, both the POFI tuna project run by the federal government from a base in Hawaii after 1947, and the Inter-American Tropical Tuna Commission investigations based at SIO and active after 1949, were designed specifically to study the entire range of biological problems associated with the tuna population in its relationship to plankton, to physical oceanographic phenomena (including temperatures, currents, and especially upwelling as a factor in nutrient supply), and to meteorological conditions. And like the sardine project, each of these two projects was given the ships, gear, and skilled work force, as well as shore laboratory facilities, to pursue the full agenda that had so ambitiously been established on an ecosystem research model.[42]

A leap of astonishing breadth was involved in this move away from the Thompson-style management-oriented research on harvests, to this sort of comprehensive ecosystem design. And so one is left to wonder how and why the ideas of Northern European and British scientists of an earlier day had seemingly sprung to life anew in this distant region. Especially intriguing is the fact that one searches in vain - either in the literature of promotion, or in the papers published out of the science that the Americans did in these Pacific projects - for more than a few scattered specific references to the work of Hjort and other leaders of the earlier ecosystem approach to fisheries studies. There was, in sum, little explicit acknowledgement of an intellectual debt to the early founders of modern fisheries oceanography.

\* \* \* \* \*

In the following paragraphs, I offer a very tentative set of clues and leads, in pursuit of an answer to this puzzle. Much of what is presented here is rather fragmentary and so constitutes only partial and incomplete evidence - but evidence worth our notice nonetheless - of lines of influence and some concrete connections. This evidence points to the conclusion that the legacy of European science from an earlier era was hardly unknown to the Americans engaged in these Pacific projects, and indeed it was certainly directly influential in the development of their grand new ecosystem research designs.

First, despite lack of specific citations in the American Pacific scientists' writings, the research contributions of the great Northern European and British students of marine ecology - not only Hjort and Graham, but among founders of the field and also their own contemporaries Gran, Petersen, Schmidt, Holt, Russell, Taning - were almost certainly studied by most, probably all, of the key figures in the design of the American postwar projects. The late William C. Herrington, who was a senior researcher under Thompson in the California marine agency and later in the international commission studies, recalled in an oral history interview, for example, that Thompson held frequent discussion groups or seminars on the literature - with his research staff and his graduate students at the University of Washington in attendance - at which the work of Hjort and other ICES researchers, as well as the British marine laboratories, was fully discussed. Herrington stated that while the Thompson-led studies had an entirely different emphasis, the European writings were well known and understood by those attending these sessions.[43]

Hjort was publishing in prominent semipopular as well as scientific journals even in the mid-1930s, moreover; hence, it is difficult to imagine that any working fisheries scientist would have been truly unaware of his thinking on issues such as the relationship of nutriment levels and juvenile survival rates to fishery abundance - not the same thing as fully developed ecosystem analysis, to be sure, but an approach that certainly went well beyond data on aggregate yields in relation to fishing effort.[44]

Second, there were a few prominent individuals who engaged in Pacific fisheries research even before 1941 whose research objectives and methodologies bore the distinctive imprint of the European and British heritage. One of them was Lionel Walford of the federal fisheries agency, who had collected larvae of sardine in the California Current aboard an SIO vessel in 1939-41 while the SIO scientists were attending to study of temperatures, chemistry and currents. Moreover, Walford's studies of the sardine stocks speculated explicitly on the impact of upwelling, temperature zones, and nutrients upon sardine reproduction and survival rates; and also upon the spatial relationship of spawning and abundance of microscopic plants.[45]

Walford's studies, because they probed the dynamics of ecological relationships in waters inhabited by the California sardine, were in the Hjort-Graham tradition. It stands as important evidence, moreover, that fisheries scientists *at the time* recognized it immediately for what it was. Thus Dr. Chapman commented in 1948 that Walford was an exemplar of the application of Northern European research methodology in American fisheries science.[46] 'In economic fisheries research in this country,' Chapman went on to say,

> there is a schism which is hard to define yet is very evident. On the one hand there are those [Walford being the most prominent] who take the broad ecological approach to fisheries problems, which derives ultimately from work which has been done in northern Europe and is excellent scientifically but has been unproductive in practical fisheries management. On the other hand there are those who strike directly at the effect of fishing mortality on the population, work back from that point, and have never been able yet to get to the point where they feel that broad ecological studies yield much [in] practical results.[47]

The last sentence in the passage quoted refers, of course, to Thompson and his school. Chapman himself was not only a student of Thompson but also had done extensive practical fisheries management research; yet he was mindful of the interesting potential of the 'broad ecological approach,' and the hopes he held out for it were evident in Chapman's own views on stressing the important need for comprehensive, ecological modes of investigation in the new Pacific fisheries projects after the war. Chapman's conviction that much greater attention needed to be given to environmental relationships was reinforced when he visited Norway in 1947 as a Guggenheim Fellow, and there came into contact with Rollefson and other marine scientists who were actively engaged in such studies in the Northern European tradition.[48]

Walford himself, moreover, entered directly the discussion among scientists in 1946-48 as to what the design and thrust of the new Pacific projects should be. He argued for 'studying normal patterns in fishery biology' in contrast to the nearly-exclusive emphasis in the commercially-oriented studies upon collection of data on fishery populations under pressure. Too much effort had been given to narrowly conceived responses to specific endangered commercial fisheries and to short-term crises. 'My view,' Walford declared,

> is that the choice is not between handling dangerous situations and being diffuse or superficial. It is rather between studying the normal mechanism of the marine organism as a whole and studying the abnormal of the separate parts....[49]

Walford thus welcomed in particular the POFI project, which would investigate a species (the Pacific tuna) that as yet had been subjected to little commercial exploitation - a species whose basic biology, environmental relationships, and population dynamics were hardly known to science.[50]

Even before the war, not only Walford (then a young scientist making his way) but also Willis Rich, professor at Stanford University and one of the senior fishery biologists on the West Coast, had begun to speak critically of the limitations of fishing-effort analysis. The California marine laboratory's statistical approach, Rich asserted in 1936, left 'other phases' insufficiently explored. He recommended that:

> much more should be done in the way of observation at sea on the spawning, growth and development of the young, [on] migrations, etc.; that the investigation should be

extended to include the fish off the coasts of Oregon and Washington and British Columbia; ... and, finally, that every possible effort should be made to discover means for determining the age and rate of growth and replacement.[51]

Rich's concern to advance a broader kind of research design was shared by others in American Pacific marine science at that time. For example, Dr. Frances Clark of the California state laboratory was already embarking on intensive study of ages in the sardine catch from the California Current. Meanwhile, the USF&WS biologists Oscar Elton Sette and Elbert Ahlstrom carried the work a step forward with regular sampling at high-seas stations off the California coast to determine changes in sardine egg volume and distribution in relation to ocean currents, over two seasons. Moreover, some serious work was being done by SIO scientists in the 'thirties, albeit mainly in a physical-oceanography mode and on a very small scale, on upwelling in relation to nutrient migration and concentration in the California Current. But still, the idea of comprehensive, ecologically-conceived studies was entirely beyond the available resources of the day, when the California marine laboratory had only a single vessel capable of deepwater studies, when only a small portion of the time at sea of two federal government vessels was devoted to such work, and when it still remained in the future for the great war-period innovations in instrumentation to effect a revolution in oceanographic techniques.[52]

In addition, several notable American marine biologists in the 1930s were taking an interest in the study of 'energy chains' and the systematic analysis of natural communities (often referred to as 'dynamic ecology'), which sought to analyze the interrelationships of physical, chemical and biological phenomena in marine communities. When the postwar Pacific projects were first being designed, for example, Dr. Carl Hubbs of SIO urged an expanded scope for proposed physiological-biochemical studies, to embrace ecosystem study.[53] It can quite well be assumed, I think, that as an active researcher in zoology and ichthyology, Hubbs must have known by then of Charles Elton's work in England - work that had given new impetus in ecology to the study of natural communities as systems - and been especially aware of Elton's analysis of food chains.[54]

More generally, because the study of limnology was a standard component of training for fisheries biologists and managers, one can also assume with a fair degree of confidence that the Pacific Coast fisheries scientists were familiar with the ecosystem style that had taken hold early in limnology and had remained the dominating paradigm of research.[55] (This was by contrast, as we have noted, with the fragmentation into physical and chemical, biological, and fisheries-management specializations of what is today a reintegrated, inter-disciplinary fisheries oceanography.)

In western Canada as well, the leading fisheries researchers were convinced, as they planned their future programs just after the war, that they needed to go well beyond harvest yield analysis. 'We have all been pressing for some time,' wrote Dr. Earle Foerster, director of the Pacific Biological Station of the Fisheries Research Board of Canada,

> for more information concerning conditions off our coast. What we need, before we can do anything for the fisheries, is a *back-log of data* on oceanographic conditions, both physical and biological, and to convince the proper authorities of this need, has been the tough job. It may be necessary to indicate clearly, and impress the fisheries [industry] people particularly, that they cannot expect early results from this work, for it may take many years before enough data are to hand on which proper correlations can be based.[56]

Further evidence that the Northern European and British tradition had inspired continuing work and new conceptions in planning by fisheries specialists, even while the Thompson school was still wholly dominating Pacific marine fisheries studies, was attested to in a 1949 memorandum to the organizers of the California CalCOFI sardine project by A.G. Huntsman, who himself had worked under Hjort's direction early in his career.[57] In that document, Huntsman discussed the difficulties he and others had encountered over many years of research - until 1934 on herring, and then afterward on salmon - in constructing a useful set of theories regarding the relationships of oceanographic and biological phenomena. Contending that 'the factors determining concentration of [marine] fish ... is an oceanographic matter,' Huntsman continued:

> It is really an ecological problem, involving the relations between organisms and their environments, between the ocean and the life therein. Twenty-five years ago I visualized it [the problem of why fish concentrate as they do] as the problem of limiting factors, of the factors limiting the distribution and abundance of marine organisms. I studied such obvious factors as temperature, salinity and light ... [but] made no particular impression. The field of study was still too vast and inchoate for easy comprehension or for solution of the problem in foreseeable time. Ecology, as being study of marine organisms and their environment,

had been immeasurably large, and even study of the *rela-tions* between organisms and their environment that determine their distribution and abundance was proving too large. How could the problem be effectively narrowed? Obvious narrowing was to take one or a few organisms and one or a few local environments. Study of an organism throughout its range in distribution seemed advisable in order to see the picture through contrasts ....[58]

All the foregoing indicates that, despite appearances on first impression that the Northern European and British heritage of ecosystem studies had been abandoned by the West Coast fisheries specialists and their institutions, this tradition was continuing to have an influence of considerable substance - even though the ships, personnel, and funding were entirely lacking until after 1947 to do much about it. We must still take account, however, of what was probably the most important single influence connecting the European tradition - connecting it palpably, and with a powerful intellectual force - with the American Pacific research efforts of the postwar era. That force was an individual and his institution: Harald Sverdrup, the Norwegian oceanographer who served as director of the Scripps Institution from 1936 to 1948 (see Plate 2).

One of the world's leading meteorologists and oceanographers, Sverdrup is properly remembered also for having organized while at SIO the first major American physical oceanography programs in the deepwater Pacific to be inaugurated since the early part of the century. He transformed the SIO teaching curriculum and research program after 1936 from one that had emphasized biological studies, to one solidly rooted in physical oceanographic studies (albeit in a diffuse way). Sverdrup also sought, however, to establish at SIO a research effort that was purposefully interdisciplinary, with attention to the full range of ecosystem relationships within ocean regions.[59] The brilliance of Sverdrup's vision in this regard was exemplified by his authoritative text, written in coauthorship with the biologist Martin Johnson and the oceanographer Richard Fleming, *The Oceans: Their Physics, Chemistry and Biology*, a work accurately described recently as 'a definitive account of the whole science, widely influential at the time and still useful for reference.'[60]

From his first months at SIO, Sverdrup sought to make the California Current a subject of study that would bring to bear the joint efforts of chemical, physical, and biological oceanographers. He had only one deepwater vessel at his disposal, the heroic little wooden schooner *E.W. Scripps*; but he put the ship to work intensively in support of his own and Fleming's pathbreaking studies of upwelling and

*Plate 2.* Harald Ulrik Sverdrup, c.1938. (Courtesy of Scripps Institute of Oceanography Archives)

currents.[61]   What is more, he welcomed the federal government's
marine biologists aboard, to work alongside the SIO oceanographers
in advancing their own work on distribution of sardine larvae in the
California Current.   Indeed, one of the principal directors of the
postwar federal projects, Dr. John Marr, argues that these collaborative
efforts on the days at sea aboard the *E.W. Scripps* before 1942 constituted the formative moment generating the ecosystem research
designs of the CalCOFI sardine project and of the other Pacific projects
after 1947.[62]

Sverdrup's extraordinary success in energizing research in physical oceanography, and simultaneously in stimulating interdisciplinary
efforts, has been persuasively established by Eric Mills.[63]   Equally
extraordinary, however, is the scope and importance of Sverdrup's
role in welcoming the Thompson-school fisheries scientists back into
the newly diversified and integrated company of oceanographers.
'Sverdrup personally didn't think much about biology,' his student
Roger Revelle recalled much later.   Nonetheless, Sverdrup certainly
appreciated the importance of connecting biological and fisheries
studies to the chemistry, physics, and meteorology of ocean science.[64]
Revelle's recollections testify to the research ambience in which he
matured as an ocean scientist at SIO.   Asked whether, as a young
geophysics graduate student working under Sverdrup at SIO in the
thirties, he had any commitment to a holistic, or ecosystem, approach
to ocean science, Revelle averred:

> I had it already - I'd always had it from my first day at sea
> in 1932.  The first work in oceanography I ever did was on
> the *E.W. Scripps*, and it involved plankton collection, plus
> measurement of the nutrients - plus temperature at salinity
> of the water.   At least at Scripps, this view was always
> there.

Although there was 'really no thought of fisheries' at Scripps
before Sverdrup took charge, Revelle stated, still a good deal of
serious work was going on in marine biology, including microbiology.
But with Sverdrup's appearance, the broad ecosystem interest became
prevalent.   The collaboration on the text *The Oceans*, Revelle argued,
should be seen as an accurate reflection of prevailing thought at SIO in
the period of Sverdrup's directorship.[65]

When a small group of West Coast fisheries specialists - led by
Wilbert Chapman (fish curator of the California Academy of Sciences),
and Milner Schaefer and Oscar Sette (scientists in the federal fisheries
agency), along with representatives of the sardine industry - came to
Sverdrup to ask for cooperation in developing a program of sardine

research in California, they found him knowledgeable, open, and enthusiastic. What was being proposed to him, after all, was precisely the kind of fisheries research that had long been the admired mode in his own country and in which Norwegian scientists had provided powerful intellectual leadership. Neither concern with the industry's problems nor joint efforts with scientists primarily interested in management - both such things being anathema, of course, to many of his fellow 'pure scientists' at SIO and elsewhere - put off Dr. Sverdrup in the least. On the contrary, he seized the occasion to reinforce Chapman's and other leaders' arguments for a broad program on an ecosystem design. In sum, Sverdrup did not betray Chapman's assessment that in addition to being a preeminent oceanographer, the SIO director had a 'broad background of training and ... the broad vision necessary to encompass the role of both physical oceanography and marine biology in high seas research', and also understood 'the viewpoint' of the commercial fisheries.[66]

The surviving manuscript records of the scientific group's earliest meetings on design of the CalCOFI sardine project reveal that Sverdrup's prestige and influence were placed solidly at the disposal of the other scientists in pushing for a program far wider in scope and more ambitious in its ecosystem focus than any ever before undertaken by Americans in Pacific waters.[67] Moreover, Sverdrup extended himself personally - taking some risks along the way, in his relationships with his university's president - when in February and March 1947 he cooperated with Revelle (then in Washington, as an officer in the U.S. Navy's Bureau of Ships) to arrange the acquisition by the University of California of three re-outfitted oceanographic vessels that would become the heart of the new Pacific scientific deepwater effort (see Plate 3).[68]

If the creative impetus for a broad ecosystem research design came from Chapman, Schaefer and others, Sverdrup thus provided an endorsement of their objectives that proved to be of critical importance. For in fact, neither the sardine project in California, in which SIO vessels and scientists would play the central role, nor the two tuna projects that expanded the postwar research effort to cover the entire Eastern and Central Pacific regions, could have succeeded without the active involvement of SIO personnel and ships - or without the exchanges of personnel, sharing of data, and use of shore laboratory facilities that SIO contributed to the interrelated programs. The infusion of federal funds and acquisition of ships that SIO enjoyed as the result of this involvement certainly vaulted the institution into new global prominence as a research center of 'big science'. Sverdrup and SIO returned more than their share, however, in giving energy and direction to the new ecosystem work at sea in the years that followed. There were major payoffs for all the partners in this enterprise.[69]

*Plate 3.* Shortly after World War II, the Scripps Institute acquired an instant fleet: *Horizon*, *Paolina-T* and *Crest* (l. to r.), to join their only earlier ship, the *E.W. Scripps*. The *Horizon* and *Crest* had been Navy vessels, the *Paolina-T* had been a fishing boat. (Courtesy of Scripps Institute of Oceanography Archives)

## II. Parallels and Patterns: The Forces that Mobilize Innovative Marine Science Projects

One of the leading historians of modern oceanographic history has written that following the long period of very limited development in marine sciences in the 1920s and 1930s, throughout the world, a phenomenal expansion of effort occurred not only in the United States but internationally after the Second World War:

> The realization by governments of the importance of marine problems and their readiness to make money available for research, the growth in the number of scientists at work, and the increasing sophistication of scientific equipment, have made it feasible to study the ocean on a scale and to a degree of complexity never attempted and never possible before.[70]

As we have noted already, the fisheries research projects promoted after 1945 by America's Pacific scientists, in league with the fishing industry and government resource agencies, were at the leading edge of this monumental expansion and flourishing of modern oceanography. What Wilbert Chapman had envisioned in his postwar campaign for new projects on the West Coast - an interrelated cluster of high seas research programs, on a scale well beyond anything previously undertaken by American science - was a dream largely realized by the U.S. programs in the Central and Eastern Pacific as early as the mid-1950s. (These projects also fulfilled Chapman's prediction in 1947 that if they could pay even small returns at first, they could attract 'within five years ... more money and support for basic ocean study than has been dreamed of'.[71]) Other countries followed as soon as they could afford it, in the 1950s and 1960s, in expansion of oceanographic programs - but the United States led; and in fisheries research, these Pacific projects led the American effort.

If I have been correct in arguing that, given the influence of the Northern European and British legacy of fisheries science, an ecosystem mode of studying marine environments was well understood by the American scientific planners in the late 1940s, we are still left to consider why they succeeded, politically and organizationally, in actually realizing the vision of comprehensive research. That is, we want to know how they mobilized the public support needed to undertake deepwater research of such vastly increased scale, intensity, and complexity. In the remaining brief section of this paper, we turn to the sources of that success in the U.S. Pacific science initiatives. Remarkable parallels are evident, even at first glance, as between the experience in Great Britain from the eighteenth century to the modern era and the experience in America after World War Two.

As Deacon has demonstrated, the promoters of ambitious scientific program designs have the difficult job of persuading their potential backers to extend the necessary patience and forebearance over the long term, while continually providing money to continue, because basic science does not always proceed on a predictable linear vector with a predictable termination point. Moreover, oceanographic work is always particularly expensive and complex; and as is true even today, the chance of an early payoff for research spending is always very remote, though promoters of ocean projects can sometimes make plausible promises of quick returns.[72]

How was it in this respect in the United States? Prior to the end of World War II, the West Coast scientists had enjoyed little success in obtaining the support they needed to do even satisfactory fisheries yield analysis, let alone basic research on marine ecosystems. Apart from the impoverishment of governments during the Great Depression, there was well-entrenched opposition from some of the major commercial fisheries, most notably the sardine industry in California, to regulation of their operations by the state government. This hostility manifested itself notably in a robust suspicion of the state's fishery scientists and their proposals for expanded research.[73]

There was little American deepwater research in oceanography generally done on Pacific waters during 1920-41, Sverdrup's studies being an exception after 1936; the small vessels, one in each state, operated by the state authorities were devoted mainly to operations in near-shore coastal waters. The federal government operated only one ship fulltime in the salmon research programs in Alaska and the waters off Washington and British Columbia; and one ship in Latin American waters, doing some scientific work but mainly exploration and experimental fishing there. Only a dozen or so scientists in all were actively engaged in Pacific marine fisheries work, and there were few doctoral students being graduated by the West Coast universities. Withal, it took a set of converging political and economic imperatives, and some truly compelling incentives for research, to turn the situation around.[74]

One such imperative was strategic and geopolitical. The United States government in September 1945 had issued what became known as the Truman Fisheries Proclamation, in which the U.S. seemed to abandon its traditional adherence to the view that waters beyond three miles were open to all nations for fishing. Now, the Proclamation declared, the U.S. stood ready to define 'conservation zones' far out to sea. Within these zones the U.S. would unilaterally or by agreement with other nations institute fishing management regulations - and even, it seemed, would be prepared to exclude other nations' fishing vessels if the U.S. deemed the fishery resource in question to be in danger of depletion.[75]

This change of fundamental policy on ocean boundaries - one that, ironically, the U.S. did not actually implement directly or indirectly for thirty years, until the 1975 extension of the exclusive fishing zone to 200 miles offshore - was seized upon as an urgent reason for major new oceanographic projects in the Pacific. Wilbert Chapman of California led the campaign, brilliantly organizing industry, science, and political groups around the argument that the U.S. had an unprecedented opportunity to find, stake out for itself, and exclude other nations from, vast Pacific fishery resources that previously had remained unexploited. Moreover, the argument was made from classical neomercantilist-geopolitical premises. In sounding this theme, Chapman declared:

> Military security follows, is implemented by, and is secured through commercial development .... If we have actively expanding fisheries in the [Pacific Rim] area, it will, in future years, reflect as beneficially upon our national security as it does upon our national economy.[76]

This kind of argument, looking toward a consolidation of America's postwar hegemony in the Pacific area by using fishery expansion as a basis for advancement of other commercial and military objectives, carried great weight in Congress when it approved large appropriations beginning in 1947, for the Hawaii-based POFI tuna research program. Soon three magnificently equipped oceanographic ships, which cooperated with the California state project and other federal ships, were working the Eastern and Central Pacific for the POFI program.[77]

The prospect of economic payoffs more narrowly defined, in terms of preserving the viability of an important commercial fishing industry, also came into play as a compelling argument for new funding and commissioning of oceanographic ships. This concerned the California sardine industry, which was experiencing a rapid drop in harvest levels, and by 1948 faced what was feared to be a 'disappearance' of sardine altogether from Northern California waters. Confronted with this crisis, the industry (traditionally opposed to regulation, or research that might lead to tightened fishing limits) now lined up with the scientists to press for state and federal financial support of the ambitiously designed sardine research program - the project which, more than any other of that day, would apply ecosystem theory on a large scale in ocean waters. As a part of the political bargain that won authorization for the program, the industry accepted a state-administered tax upon its own fish landings. Over the next twenty years, even after the commercial sardine fishery industry had completely collapsed in California because of depletion, the special tax continued on other species that were landed; and it supported a large component of the comprehensive CalCOFI program of research on the California Current ecosystem.[78]

As Deacon has stated, sometimes scientists 'pretend' to do applied fisheries studies (as she suggests happened at Plymouth) in order to get the public support required for long-term efforts at *basic* research.[79] In the case of the California sardine project, the scientists pursued a research design which (precisely because it sought to embrace the ecosystem) was ever-ramifying; and no new harvest regulations or other policy proposals, other than their proposals for expanded research, were endorsed by the project leaders. Indeed, an eminent West Coast fisheries scientist privately ventured the 'rather unkind' observation that the CalCOFI project was most obviously 'a tremendous windfall to the Scripps Institution'. Beyond that, he wrote, its objectives seemed to be '[affording] employment to large numbers of fisheries biologists and oceanographers', and giving the sardine industry an excuse to delay the advent of new regulations. The narrower type of harvest studies that he thought were truly necessary, for meaningful support of a program of 'conservation regulations', were not being vigorously pursued.[80]

The CalCOFI scientists asserted, however, it was their role to conduct research and not to make specific policy recommendations. The sardine industry accepted the scientists' view with an almost perverse enthusiasm - a Faustian bargain as it were, the price of their entrepreneurial freedom being their industry's imminent economic collapse. Meanwhile the California fleets engaged in fishing for species other than sardine continued to finance the research project - led in this respect by a few industry leaders who regarded the ever-expanding research effort as the indispensable foundation of 'sane conservation' in the longer run.[81]

The first tuna project (POFI) had a different kind of organizational and political dynamic. It was supported by the tuna fishermen, though with some apprehensiveness about regulation, because its initial emphasis was on locating previously undiscovered tuna stocks; regulation of any kind was far in the future. The industry gave its full support to the basic science component of the work, and it was richly rewarded with discovery of what quickly became the fastest growing and most profitable fishery in the Eastern Pacific. Such spectacular practical success helped to maintain the support enjoyed by the project in Congress, and the appropriations kept flowing at generous levels well into the 1960s. Meanwhile, the project had widely publicized spin-off benefits in pure science, including important discoveries relating to Eastern Pacific currents.[82]

The third of the major new initiatives was the Central and South American tuna investigation, under the agreements with Costa Rica and Mexico in 1949. This research, supervised by the brilliant fisheries biologist Milner Schaefer, fulfilled the high hopes of those who had predicted important results from ecosystem study of species in a 'normal' state, rather than under intensive exploitation. From this work Schaefer developed his influential new models of fishery population dynamics; and the project became a showcase of modern fisheries oceanographic scholarship, while also setting the stage for an international regulatory agreement and quota limitations in the mid-1960s.[83] The U.S. tuna industry, for its part, showed initial reluctance to support this initiative. Thus Wilbert Chapman - the fisheries scientist and research promoter of the 1947-49 projects, now a State Department officer in charge of the new treaty - was frustrated in 1948 by the tuna industry's foot-dragging despite the remoteness of regulation. 'I feel moderately sure,' he despaired, 'that the industry would rather face the prospect of canning orange juice and soup before it would admit the desirability of any regulation whatever'. But finally the industry did throw its weight mightily into the balance in the political process of obtaining appropriations for vessels, operations and laboratory work.[84] They did so because the State Department assured them that a fund of research data would be a hedge against unilateralist moves in the future by the Latin American states, claiming endangerment to 'overfished' stocks, to exclude U.S. vessels from their waters.[85]

Further impetus for support of the new programs came from the U.S. Navy, which had its own interest in promoting study of Pacific oceanography. This reflected the norm in all nations that were naval powers: the military uses of oceanographic research were of vital importance, and naval leaders naturally encouraged new initiatives and hoped to use them for their own purposes.[86] For many years Navy ships had been collecting data on currents and depth soundings in the Pacific, depositing at SIO the information for archiving and processing. The Hydrographic Office of the Navy gave enthusiastic support in 1946-47 to plans for the sardine project, anticipating the benefits of data from the vast grid of stations this project would establish. The Navy welcomed the prospects of 'extensive synoptic oceanographic information about the waters off the Pacific Coast', and especially of 'expanded investigations of the departures from normal oceanic circulation'.[87] It was at the Navy's initiative, moreover, that three warships were designated by Congress to be converted to oceanographic uses, outfitted, and turned over as a gift to SIO.[88]

Finally, as we learn from the history of large-scale British and European scientific projects, typically it is difficult to obtain governmental funding commitments for anything like the very long period that such projects actually require. (Even the greatest British geographic surveys, as Deacon reminds us, were 'sold' to Parliament on the basis of only five-year plans and appropriations requests.) Few promoters of large-scale projects in basic science have had the temerity to admit candidly, as did the promoter of America's great Wilkes Expedition of 1838-42, that by an 'immutable law of nature' that applied to 'all matters of science ... utility cannot be computed in advance', and so to a great degree government patronage must be advanced on faith alone.[89] To the credit of those who pressed for the Pacific Ocean projects chronicled here, they did sound a warning; thus Wilbert Chapman told a leading industry figure, as he also testified before Congress, in 1947, that basic biological research - essential for rational regulation of a fishery, 'is expensive and time-consuming'.[90]

Once a major scientific project does begin to produce dramatic results, however, it can expand its base of support from economic interest groups or (as often is equally important) the weight of respected opinion in the scientific establishment. This is precisely what happened with the tuna projects (i.e. POFI and Schaefer's research program). The dynamics of support for the CalCOFI project, as sardine disappeared and the program's emphasis became progressively broader, were rather different. Here, perhaps, one can credit above all the rising public interest more generally in support of oceanography, by the mid-fifties; and, particular to California, the fact that appropriations to SIO were built into the budgets of a rapidly expanding University of California research and teaching system. Moreover, the very breadth of the CalCOFI research made it attractive for the Navy and later private ocean industry firms to give grants and contracts to SIO, and so buy into the data base and expertise that the CalCOFI studies had been building up within its expanding faculty and research staff since 1948. Hence, along with the University of Washington, which had similarly been advantaged by its rich background of involvement in fisheries research, SIO became a leading participant in (and beneficiary of) the major upswing in governmental support of oceanographic research in the 1960s and 1970s. Indeed, not until the fiscal crisis in America in the early 1990s did the indicators of expansion of SIO research programs and proliferation of governmental and private support to the institution show signs of possibly serious long-term reversal.[91]

An inventory of the historic significance of these American research efforts in Pacific waters would have to include, withal, the way in which they inaugurated the beginning of the movement that established oceanography as 'big science' in America - with research organized in large institutions, interdisciplinary teams dominating the field in place of the individual researcher, and reliance on steady large-scale governmental support for ships, instruments, and training of personnel.[92]

Another major achievement of the era was the process, painstakingly planned and devotedly nurtured by the leading Pacific Coast fisheries oceanographers, by which the oceangoing research efforts of the new projects were coordinated and integrated. In the process, they tested and proved the exciting possibilities that inhered in application of recent advances in instrumentation - especially sonar, but also including radar, aerial photography, improved bathythermography, and, at least putatively, computing techniques for the management of large synoptic data collections from the grid stations that CalCOFI established. Virtually the whole spectrum of techniques and interdisplinary modes of investigation associated with current-day fisheries oceanography was advanced by their applications in the postwar American projects in Pacific deepwater regions.[93]

An additional tangible result of these projects was the California sardine program's creation of the enormous grid that has provided the longest-term set of systematic sampling of ocean chemistry, physics, and flora and fauna that is extant today. The sudden burst of concern for data relevant to measurement of global warming's possible effects has given this body of California Current data a unique importance for the study of global climate change.[94]

Viewed from today's vantage point, half a century later, the most luminous achievement of the Pacific projects is surely their record of advancing holistic studies of marine environments. Their research - which pursued the analysis of processes of change in complex biotic communities, instead of studying segmented processes of very small deepwater spatial areas - anticipated by a good fifteen years the ecosystem mode of studies that won acceptance much more broadly in the 1960s, when the ecological paradigm re-emerged so powerfully in natural science. The latter development also represented, of course, the triumph of ideas that had been pursued in the earliest marine ecological research in Northern Europe and Great Britain. America's postwar Pacific Ocean research served as an intermediary and catalyst in that process, one of the great transforming episodes in the history of modern ocean science.

## NOTES

1    This paper is funded by a grant from the National Sea Grant Col-
     lege Program, National Oceanic and Atmospheric Administra-
     tion, U.S. Department of Commerce, under grant number
     NA89AA-D-SG138, project number R/MA-30 through the
     California Sea Grant College, and in part by the California State
     Resources Agency. The views expressed herein are those of the
     author and do not necessarily reflect the views of NOAA or any
     of its sub-agencies. The U.S. Government is authorized to
     reproduce and distribute for governmental purposes.
     The author is grateful for suggestions and comments on the
     research while in progress to his colleagues Roger Hahn and John
     Dwyer, University of California, Berkeley; Eric Mills, Dalhousie
     University; James Sullivan, California Sea Grant College
     Program; Deborah Day, archivist of the Scripps Institution of
     Oceanography; Janet Ness, University of Washington Libraries;
     John Marr, U.S. Fish and Wildlife Service (ret.); Margaret Deacon,
     Southampton University; and Walter Lenz, University of Ham-
     burg. David Williams, Stephen Fisher, and other participants in
     the 1991 Exeter University Dartington Conference offered highly
     useful suggestions in discussions of the conference draft at
     Dartington.
2    Harry N. Scheiber, 'Wilbert Chapman and the revolution in U.S.
     Pacific ocean science and policy, 1945-1951', in R.F. MacLeod and
     P.F. Rehbock, *Nature in Its Greatest Extent: Western Science in the
     Pacific* (Honolulu, 1988), pp. 223-44. See also Scheiber, 'Pacific
     ocean resources, science, and law of the sea', *Ecology Law
     Quarterly*, 13 (1986), 383-534.
3    'CalCOFI' is the acronym of the California Cooperative Fisheries
     Investigation. In its founding period, it was known as the Marine
     Research Committee. See Harry N. Scheiber, 'California marine
     research and the founding of Modern fisheries oceanography:
     CalCOFI's early years, 1947-64', *CalCOFI* (California Cooperative
     Fisheries Investigation) *Reports*, 1990, 53-74; and Arthur McEvoy
     and Harry N. Scheiber, 'Scientists, entrepreneurs, and the policy
     process: A study of the post-1945 California sardine depletion',
     *Journal of Economic History*, 14 (1984), 393-413.
4    First congressional appropriations for POFI came in 1947, and the
     work was in full swing by 1948. See O.E. Sette, *Progress in Pacific
     Oceanic Fishery Investigations*, 1950-53 (U.S. Department of the
     Interior, Special Scientific Report: Fisheries, No. 116) (Washing-
     ton, 1954). On the POFI record, see note 82 below.
5    Milner B. Schaefer, 'Management of the American Pacific tuna
     fishery,' in Norman Benson, ed., *A Century of Fisheries in North*

*America*, (Washington, 1970) 237-48. See also note 6, below. Other new Pacific projects of the post-war era, it should be noted, were primarily concerned with physical oceanography; notable among them were EPOC and NORPAC.

6    I have explored in another paper the issue of how these interrelated oceanographic studies were also designed to advance American security and commercial expansionist aims: cf. Scheiber, 'U.S. Pacific fishery studies, 1945-1970: oceanography, geopolitics, and marine fisheries expansion', in Walter Lenz and Margaret Deacon, eds, *The Ocean Sciences. Their History and Relation to Man* (Hamburg, 1990), 471-521. For full background on the history of fisheries science and its relation to policy in California, see Arthur F. McEvoy, *The Fisherman's Problem: Law and Ecology in the California Fisheries, 1850-1980* (Cambridge, 1986); and on fisheries science issues in the immediate postwar period, Scheiber, 'California marine research', 63-70. Integrating marine biology with physical oceanography was not, of course, the modal style in fisheries research and management in Northern Europe. Rather, it was the subject of limited experimentation in a few agencies and marine laboratories, mainly in Norway, Denmark, and the UK. And even in those centers of study, such ideas had generally been applied to problems of limited scale in ocean areas, or to fisheries that had been under intensive study for decades. See, on British research, Charles E. Lane, 'Biology of the sea', in C.P. Idyll, ed., *Exploring the Ocean World: A History of Oceanography* (New York, 1969), pp.42 ff.

7    The ecological tradition can be dated in marine science, of course, at least from Edward Forbes' work in the 1840s; Karl Möbius, Ernst Haekel, Victor Hensen, and H.H. Gran all undertook systematic study of plant or animal species in relation to the marine environment. See the concise overview and analysis in Frank N. Egerton, 'Ecological studies and observations before 1900', in Benjamin J. Taylor and T.J. White, eds, *Issues and Ideas in America* (Norman, Oklahoma, 1976), 339 ff.

8    Margaret Deacon, *Scientists and the Sea, 1650-1900: A Study of Marine Science* (New York, 1971); and Eric Mills, *Biological Oceanography: An Early History, 1870-1960* (Ithaca, New York, 1989).

9    Eric L. Mills, 'The ocean regarded as a pasture: Kiel, Plymouth and the explanation of the marine plankton cycle', in Lenz and Deacon, eds, *Ocean Sciences*, 20-9, for analysis of two such episodes with varying results and degrees of success, each of the initiatives in question (in the field of biological oceanography) organized differently and led by men of varying temperaments and intellectual styles; and Mills, 'The oceanography of the

Pacific: George F. McEwen, H.U. Sverdrup and the origin of physical oceanography on the West Coast of North America', *Annals of Science*, 48 (1991), 261-5, for analysis of such a moment of new initiative and success in physical oceanography in Pacific offshore waters - and for discussion also of the dynamics of false starts, partial successes, and frustrations. The author is much indebted to Dr. Mills' insightful writings on the historical inter-pretive problems regarding both the internal history and the social context of oceanographic science.

10     On Hjort's brilliant conceptualization of marine fisheries research and its influence on the Continent and in Great Britain, see E.S. Russell, *The Overfishing Problem* (Cambridge, 1942). As will be noted below, early limnology was also pursued as an ecosystem model.

11     See, *inter alia*, David Cushing, *Fisheries Resources of the Sea and Their Management* (Oxford, 1975). See also M. Sinclair and P. Solemdal, 'The development of a "Population Thinking" in fisheries biology between 1878 and 1930', *International Council for the Exploration of the Sea, Biological Oceanography Committee*, No. C. M. 1987/L:11 (mimeographed, Canada Department of Fisheries and Oceans, Halifax Fisheries Research Laboratory).

12     Robert P. McIntosh, *The Background of Ecology: Concept and Theory* (Cambridge, 1985), 49-56. For the evidence of the explicit recog-nition of the need to be concerned, in early fisheries science research, with habitat and ecosystem relations as an essential basis for management decisions, *cf.* discussion of the 1880s Scot-tish debates, in Margaret Deacon, 'The establishment of marine stations and early commercial fisheries research in Scotland', in Harry N. Scheiber, ed., *Ocean Resources: Industries and Rivalries Since 1800* (Center for the Study of Law and Society, University of California Berkeley, Working Papers, 1990); see also Dr. Deacon's contribution to the present volume from the 1991 Dartington Conference.

13     McIntosh, *Background of Ecology*, 115. But *cf.* text following note 24 below.

14     This is not to say that ecological marine science, with respect to the major concern with plankton studies, was a unified field without internal divisions of emphasis and methodology. See the full discussion in Mills, *Biological Oceanography, passim*.

15     Reference is to the study directed by George Clarke, zoologist, and Gordon Riley, marine physiologist, which provided from an ambitious grid of collecting stations at sea the data for Clarke's analysis of the energy flow and food chain that occurred in the ecosystem on the banks. These data later provided the empirical basis for the path-breaking work of Henry Stommel and his col-

leagues, in developing a theoretical model of the marine food chain and its dynamic energy 'budget.' The Georges Bank study is treated fully in Mills, *Biological Oceanography*, 273-81; and also is discussed in Susan Schlee, *On Almost Any Wind: the Saga of the Oceanographic Research Vessel Atlantis* (Ithaca, N.Y., 1978), 86-92.

16  J.L. McHugh, 'Trends in fishery research', in Benson, ed., *A Century of Fisheries*, 25-56.

17  *Ibid.*; McEvoy, *The Fisherman's Problem, passim. Cf.* Cushing, *Fisheries Resources*, for discussion and critique of this yield analysis approach; see also note 23, below.

18  The phrase is now used routinely by fisheries investigators who engage in ecosystem studies. For the shift from the old to new style, see Scheiber, 'California marine research', 63-84.

19  W.A. Carrothers, *The British Columbia Fisheries* (Vancouver, 1941). See also Frank N. Egerton, *Overfishing or Pollution? Case History of a Controversy on the Great Lakes* (Great Lakes Fishery Commission, Technical Reports, No. 41, January 1985) for discussion of application of fishery yield analysis to Great Lakes depletion concerns.

20  See brief review of Thompson's work and ties with industry in 'Fisheries research institute marks cooperation between industry and University of Washington', *Pacific Fisherman*, July 1947, 39-40. See also Larry Nielsen, 'The evolution of fisheries management philosophy', *Marine Fisheries Review*,38 (1976), 15-23.

21  Ironically, one of Thompson's earliest contributions to the halibut research effort did not deal with harvest analysis at all, but rather was in the realm of basic science: he discovered that annual rings on the otolith of the halibut were evidence of the age of the fish. In later years, he abandoned any emphasis on age determination. See Cushing, *Fisheries Resource*, 47-8.

22  Indeed, as will be discussed further below, there was a shift toward more distinctly ecological-style investigation in some of Thompson's work in the mid-thirties; see, e.g. Thompson and R. Van Cleve, 'Early life history of the Pacific halibut, 2: Distribution and early life history', *Reports of the International Fisheries Commission*, 9 (1936), 1-184.

23  See Thompson's classic explications of the yield-analysis method in Thompson, 'Theory of the effect of fishing on the stock of halibut', *Reports of the International Fisheries Commission, No. 12* (Seattle, 1937); and Thompson and F.H. Bell, 'Biological statistics of the Pacific halibut fishery. II: Effect of changes in intensity upon total yield and yield per unit of gear', *ibid.*, No. 8 (Seattle, 1934).
    For an overview and analysis of Thompson's methods, see also Russell, *The Overfishing Problem*, 75-101; and, for a succinct modern critique of this approach, rejecting statements from the Thompson

fishing-effort model as often 'no more than simple arithmetic by analogy', G.L. Kesteven, 'Management of the exploitation of fishery resources', in Brian J. Rothschild, ed., *World Fisheries Policy: Multidisciplinary Views*, (Seattle and London, 1972), 237 and 229-62, *passim*.

24 Mills, *Biological Oceanography*, p.4. Mills here takes issue specifically with the contention of McIntosh, quoted above in the text at note 13.

25 Among these many students most notable were Wilbert Chapman, principal organizer of the California sardine project (CalCOFI) in 1947-8; Richard Van Cleve, head of research for the California marine laboratory and later Thompson's and Chapman's successor as director of the School of Fisheries in the University of Washington; Milner (Benny) Schaefer, first director of research for the federal government's Hawaii-based tuna project (POFI) and later first director of the Inter-American Tropical Tuna Commission research program. William C. Herrington, head of the Japan fisheries regimes under MacArthur's Occupation administration and later a State Department officer in charge of international fisheries issues, was a research associate in the halibut commission work under Thompson.

26 These observations are based upon the author's study of the Thompson correspondence and writings, William F. Thompson Papers, University of Washington Archives.

27 Thompson to Chapman, 11 June 1946, Thompson Papers, University of Washington Archives.

28 *Ibid.* 'There will be many years of exploration, accumulation of background data, and expansion of the commercial fisheries' in the tuna waters of the Pacific, Thompson cautioned, before the goals that Chapman so 'light-heartedly' (as Thompson said) was proposing.

29 Author's interview of Roger R.R. Revelle, La Jolla, 12 March 1986. It should be noted that the Friday Harbor laboratory was visited by H.H. Gran in 1928 and 1930; he introduced at that time phytoplankton research on the European model. See K.R. Benson, 'H.H. Gran and the development of phytoplankton research on the American West Coast', in Lenz and Deacon, eds. *Ocean Sciences*, 375-7.

30 Wilbert M. Chapman to Miller Freeman, 11 August 1947, Miller Freeman Papers, University of Washington Libraries.

31 Revelle interview, 12 March 1986.

32 *Ibid.*

33 See citations at notes 2 and 3 above.

34 Donald Fleming, 'Roots of the new conservation movement', *Perspectives in American History*, 6 (1971), 7-91.

35 'Memorandum on the need for oceanographic studies for Pacific coast fisheries', MS., marked 9 October 1946 (signed by Joseph Craig, Frances N. Clark, Donald McKernan, and Oscar E. Sette), copy in Director's Files, SIO Archives, University of California, San Diego.

36 Roger Revelle to Col. I.M. Isaacs, 29 November 1947, SIO Archives; see also Revelle's memorandum on the sardine project, 3 May 1948, manuscript in Subject Files: Marine Life Research Program, SIO Archives.

37 For a full exposition of this interpretation, see Scheiber, 'California marine research', *loc. cit.*

38 Revelle to Isaacs, 29 November 1947, SIO Archives, *loc. cit.*

39 How intensive the study and extensive the area is indicated by the map of collecting stations at which the project's ships took chemical and biological samples, and did physical readings, on a monthly basis over a grid stretching from British Columbia to Baja California waters, and 400 miles out to sea - over 670,00 square miles in all, with stations only 120 miles apart. See map of the grid in Scheiber, 'California marine research', p. 71. See also, *inter alia*, McEvoy, *Fisherman's Problem*.

40 Revelle, remarks to the conference on 'The position of SIO in the university, the state, and the nation', La Jolla, California, 1951, transcript of proceedings, copy in SIO Archives. On similar lines, Wilbert Chapman, the chief promoter of the sardine project, viewed it as only one strand in a 'web of research' that would embrace the entire Pacific Ocean. (Chapman, Statement in testimony, in U.S. 80th Congress, 1st Session, House of Rep. Comm. on Merchant Marine and Fisheries, *Development of High Seas Fishing Industry: Hearings*, 12 May 1947 (Washington, 1949)).

41 See A.E.J. Went, 'Seventy years a'growing: A history of the International Council for the Exploration of the Sea, 1902-1972', ICES, *Rapports et proces-verbaux des Reunions*, 165 (1972), 1-252; and J.R. Dymond, 'European studies of the populations of marine fisheries', *Bulletin of the Bingham Oceanographic Collection*, 11 (1948), ART. 4, 55-80. As Susan Schlee has observed, even within ICES hydrographic scientists and fisheries biologists were soon pushing their research in distinctive (and diverging) directions: 'The two groups resisted hyphenation, and each proceeded to work out its own story without being greatly influenced by the other'. Schlee, *Edge of an Unfamiliar World: A History of Oceanography* (New York, 1973), 212.

42 On the joint inter-American research program, see Milner Schaefer, 'Scientific investigation of the tropical tuna resources of the Eastern Pacific', in *Papers Presented at the International Technical Conference on the Conservation of the Living Resources of the Sea*

(Rome, 1955), 194-222. The research record of POFI will be examined fully in a forthcoming paper by the present author, but the program is summarized in Sette, *Progress in Pacific Oceanic Fishery Investigations, loc. cit.*

43   Author's interview of William C. Herrington, Connecticut, June 1988. This assertion does not necessarily contradict, however, the conclusion of Keith R. Benson that in the 1920s there was an 'almost complete lack of appreciation for the work of Victor Hensen and Karl Brandt,' who had pioneered in ecological studies centering on the plankton and their environment. At least this element of ecosystem study, however, came into American Pacific research when Gran visited the University of Washington laboratory in 1928 and 1930, and especially when Martin Johnson at SIO pushed forward with studies that gave attention to dynamic relationships in marine biology. (Benson, 'H.H. Gran and the development of pytoplankton research', in Lenz and Deacon, eds, *Ocean Sciences*) Herrington himself left the West Coast in the mid-1930s to conduct investigations of haddock fishing as affected by gear regulation in the North Atlantic. Johnson and Gran were associated in studies of Passamaquoddy Bay in 1932. (Eric Mills has graciously provided this last reference.)

44   See Hjort, 'Human activities and the study of life in the sea', *Scientific Monthly*, 25 (October 1935), 630-64. The California state marine laboratory scientists had for some years been tagging and also conducting age-group analysis based on laboratory study of catch samples; but as the former director of the laboratory recalled in 1949, they had discontinued a program of study on juvenile sardine, apparently because of the difficulty of finding schools of the smaller fish - a problem that was partially solved after 1947 by the application of sonar and use of newly developed types of collecting gear. (Richard Van Cleve to Frances Clark, 29 December 1949, Van Cleve Papers, University of Washington Archives.) See 58, below.

45   Walford, 'Correlation between fluctuations in abundance of the Pacific sardine (*Sardinops caerulea*) and salinity of sea water', *Journal of Marine Research*, 6 (1946), 48-53; Harald Sverdrup to Charles Hatcher, 23 February 1948, SIO Archives (referring to the importance Walford's biological and ecological studies). Earlier, it is no surprise to find, Walford had done studies in the North Atlantic on the relationship of the physical ocean environment to haddock egg and larvae migration and survival. (U.S. Bureau of Fisheries, Bulletin, 49 (1938)). And in 1939 Walford had urged the federal Bureau of Fisheries to support research on the general problem of relationships between currents and abundance fluctuations. (Monthly Reports, October 1939, MSS., Box 485A, File

825.2, US Fish and Wildlife Service Records, Record Group 22, National Archives, Washington.)

46    Chapman to Carl Hubbs, 30 April 1948, copy in Chapman Papers, University of Washington Libraries.

47    *Ibid.*

48    Chapman's views and career are fully analysed in Scheiber, 'Pacific ocean resources', *loc. cit.*

49    Walford, 'The case for studying the normal patterns in fishery biology', *Journal of Marine Research*, 7 (1948), 507, 510.

50    *Ibid.*, 510. Note how Revelle too had stressed the importance of shifting attention from equilibrium analyses to dynamic analysis, quotation in text at note 36 above. E.S. Russell observed in 1942 that it is 'only rarely [that] fishery scientists have the chance to study stocks that have not been subject to man's exploitation through commercial fishing'. That the age distribution of such fishery stocks would vary enormously, with much higher propor-tion of mature fish, from those under exploitation, had been learned from the discovery of a formerly unknown plaice grounds off the northeast coast of Norway in 1907. (Russell, *Overfishing Problem*, 1, 77.) It was just such a rare opportunity that Walford anticipated would make the POFI tuna project so impor-tant.

51    Willis Rich to Elmer Higgins, 30 April 1936, Box 486, File 825.9, USBCF General Files, USF&WS Records, Record Group 22, National Archives. Rich had also done important studies, earlier in the century, on the salmon; and in the immediate postwar period he was to serve briefly as an adviser to the Occupation on fisheries policy in Japan.

52    On Frances Clark's age-class studies and their contribution, see her landmark study, 'Measures of abundance of the sardine (*Sardinops caerulea*) in California waters', California Fish & Game Division, *Fish Bulletin*, No. 41 (1939); see also discussion of Clark and her colleagues in McEvoy, *Fisherman's Problem, passim*. The federal research team's study (U.S. Fish and Wildlife Service) was reported in Sette and Ahlstrom, 'Estimations of abundance of the eggs of the Pacific pilchard (*Sardinops caerulea*) off Southern California during 1940 and 1941', *Journal of Marine Research*, 7 (1948), 511-42. See also note 43, *supra*, on Sverdrup's interest in advancing Walford's early research on sardine spawning in rela-tion to chemistry of the ocean.

53    Hubbs, Memorandum (MS.), 15 July 1947, 'Desirable expansions in the biochemical-physiological program', Carl Hubbs Papers, SIO Archives.

54    Elton, *Animal Ecology* (1927); see discussion in Donald Worster, *Nature's Economy: The Roots of Ecology* (San Francisco), 294-306. It

must be noted, however, that Hubbs' personal index-card biblio-
graphical files had only two citations (1933 and 1938) to Elton's
works; and there are no references to Elton in the index to Hubbs'
writings. Frances Hubbs Miller, *The Scientific Publications of Carl
Leavitt Hubbs: Bibliography and Index, 1915-1981* (Hubbs-Sea World
Research Institute, Special Publication No. 1), (San Diego, n.d.). I
am grateful to Mrs. Betty Shor, the SIO historian, La Jolla, for her
advice in regard to Hubbs' scholarship.

55    This view of the importance of limnology's relationship to the
new ocean studies is supported by the analysis in Joel Hedgpeth,
'Concepts of marine ecology', *Treatise on Marine Ecology and
Paleocology*, 29-52. See also the insightful discussion in McIntosh,
*Background of Ecology*, 57-61, 119-27; McIntosh, 'Ecology since
1900', in Taylor and White, eds, *Issues and Ideas in America*, 355;
and Egerton, 'Ecological studies', *ibid.*, 339-40.
Worster's work, *Nature's Economy*, is deficient in its lack of any
attention to the marine biological tradition in fisheries from the
time of Hjort's first major contributions to the post-World War II
era, as it also neglects the impact of the limnologists from the
1920s on fisheries science. The fisheries studies in an ecosystem
mode, both of the early twentieth century and the modern era
whose opening phase after World War II is discussed in this
paper, are also notably absent from the historical reconstruction
attempted in Anna Bramwell, *Ecology in the Twentieth Century: A
History* (New Haven and London, 1989).
Prominent among the European scientists who had early sought
to apply in ocean regions some of the limnological paradigm was
C.G. Johannes Petersen, who with his colleagues traced relationships among environmental conditions, food supply, and fishery
population dynamics in the Limfjord, Denmark. See Petersen,
'The sea bottom and its production of fish food', *Reports of the
Danish Biological Station*, (1918); and discussion in Herdman,
*Founders of Oceanography*, 273-5, 324-7; and Susan Schlee, *Edge of
an Unfamiliar World*, 21-417. Also of seminal importance, in this
tradition, was Sir Alistair Hardy's investigation in the mid-1920s
of the biology of the whale in relation to the Antarctic ecosystem.
(See Idyll, 'Science of the sea', 50-2).

56    R.E. Foerster to Wilbert Chapman, 25 March 1947, Chapman
Papers, University of Washington Libraries.

57    As Eric Mills has called to my attention, Huntsman assisted Hjort
when the latter was employed by the Canadian Government to
an oceanographic survey of the Scotian Shelf and Gulf of St.
Lawrence in 1915. See Mills, 'Doing the right science at the
wrong time: The Canadian fisheries expedition of 1914-15',
(Abstract of Lecture, Madison, Wisconsin, November 1991).

58   Huntsman, 'Fisheries oceanography', MS. (marked as received in May 1949 by the SIO Director's Office and circulated at SIO), SIO Subject Files: Marine Life Research, SIO Archives).

59   Sverdrup's role in transforming not only SIO but Pacific ocean science, in this regard at least, is treated most insightfully by Eric Mills in 'The oceanography of the Pacific,' 261-6. See also Mills, 'Useful in many capacities', 297, quoting Sverdrup on his decision to begin integrated research in a small area, given the impossibility (for lack of resources) of undertaking a survey of the Pacific on a grander scale: '[It] has been my hope, since I became acquainted with the Scripps Institution ... to place emphasis on what all the special fields have in common - the ocean itself - in order to develop a program for the Institution as such, in which as many as possible of our specialists can participate. In this manner results in one field may find application in another, and more rapid advances may be expected'. Sverdrup, 'Research in oceanography: California scientists study the ocean', *California Monthly*, December 1940, 11.

60   Mills, 'Oceanography of the Pacific', 263.

61   *Ibid.*, 264; see also, *inter alia*, Shor, *Scripps Institution.*

62   Telephone interview with Dr John Marr, 19 March 1990.

63   Mills, works cited in notes 8-9, *supra.*

64   Author's interview of Revelle. This is borne out, in a fairly strong way, by Sverdrup's admission to California scientists and industry people that he had no information of herring or sardine 'disappearances' in Norway's offshore waters, and that he would need to write to his colleagues at home to learn of what studies had been done. Sverdrup, Correspondence with Wilbert Chapman and others, 1947, SIO Directors' files, SIO Archives.

65   Revelle interview.

66   Chapman to Vernon Knudsen, 22 October 1947, Chapman papers, University of Washington Libraries.

67   'Notes of Meeting in San Francisco on January 24, 1947', Carl Hubbs Papers, SIO Archives.

68   See Shor, *Scripps Institution*, 37*ff.* The University of California, thanks to Revelle's ingenuity, had - as President Sproul ironically put it - acquired its very own navy overnight; Sproul did say, not without some anger, that he would have preferred to have been consulted in advance about the negotiations! But after a timely visit from a sardine industry delegation, Sproul came around and gave his retroactive blessing to the new fleet. (Sverdrup-Sproul correspondence, February-March, 1947, Directors' Files, SIO Archives; SIO files, President's Records, University of California, Berkeley, Archives). See also McEvoy and Scheiber, 'Scientists, entrepreneurs', 401-2.

69 McEvoy argues, however, that the payoff in the California project case, was not so good for the fish. This contention rests on the view that the scientists did not, as they should have done, advocate strong regulatory controls on the overfishing that was endangering the sardine population - and that by the 1950s had seemingly destroyed the fishery, which only forty years later has begun to come back to commercially harvestable levels. See McEvoy, *Fisherman's Problem*; and, for a quite different judgment on the role of the scientists in this matter, Scheiber, 'California marine research', 74-5.

70 Deacon, *Scientists and the Sea*, 394.

71 Chapman to Carl Hubbs, 13 August 1947, Subject Files, SIO Archives.

72 Where not otherwise indicated, my references to Deacon's arguments are to her paper in the present volume. But see also her discussion of the Scottish fisheries projects and their origins, in 'State support', in Scheiber, *Ocean Resources*; and her study of the pre-nineteenth century history in her classic work, *Scientists and the Sea*.

73 The federal government's fishery scientists tended to align themselves with industry in this regard - according to one student of those years, because they probably were fearful of political consequences if they did not. A.F. McEvoy, 'Law, public policy, and industrialization in the California fisheries, 1900-1925', *Business History Review*, 57 (1983), 494-521; *cf.* McEvoy and Scheiber, 'Scientists, entrepreneurs, and the policy process', 393-7.

74 Scheiber, 'California marine research', 65-7; Shor, *Scripps Institution*; Mills, 'Oceanography of the Pacific', 265-6.

75 On the Truman Proclamation and its tumultuous early history, see Anne Hollick, *U.S. Foreign Policy and the Law of the Sea* (Princeton, 1981), 18-126.

76 Chapman, circular letter to all Congressional representatives from the West Coast and others, 25 September 1945, Chapman Papers, University of Washington.

77 See Scheiber, 'U.S. Pacific fishery studies', in Lenz and Deacon, eds, *Ocean Sciences*.

78 Documentation on the sardine project may be found in Scheiber, 'California marine research', and in McEvoy, *Fisherman's Problem*.

79 Deacon, presentation at the 1991 Dartington conference.

80 Richard Van Cleve to Frances Clark, 13 August 1951, Van Cleve Papers, University of Washington Archives.

81 So explained by Phister, a tuna packing company executive, in a letter to Van Cleve, 5 December 1949, Van Cleve Papers, University of Washington Archives.

82 Schaefer, 'Pacific tuna fishery', 237-42; Robert C. Cowen, *Frontiers*

*of the Sea: The Story of Oceanographic Exploration* (Garden City, New York, 1969), 165-6 (on POFI researchers' discovery of the Cromwell Current).

83  For a summary of studies, see *Scientific Investigation of the Tropical Tuna Resources of the Eastern Pacific*, in United Nations Document a/CONF.10/L.1, pp. 194-222; and for discussion, Cushing, *Fisheries Resources*, 27-8, 58-61.

84  Chapman to Richard Croker, n.d. (marked 1948 or 1949, but 19 September 1948), copy in Chapman Papers, University of Washington Libraries.

85  Chapman to tuna industry leaders, 25 August 1950, copy in Chapman Papers, University of Washington Libraries.

86  For such an interrelationship of science and the navy in Germany, see Walter Lenz's case study of the interwar period, in Lenz and Deacon, eds, *Ocean Sciences*.

87  R.O. Clover (Navy Hydrographic Office) to Albert M. Day (Fish and Wildlife Service), 11 February 1947, copy in SIO Directors' Files, SIO Archives. See Mills, 'Oceanography of the Pacific', 248-50, 253-4, on the Navy's activities and earlier ties with SIO.

88  *Supra*, note 66.

89  Jeremiah Reynolds, quoted in William Stanton, *The Great United States Exploring Expedition of 1838-1842* (Berkeley and Los Angeles, 1975), 31.

90  Chapman to Montgomery Phister, 2 May 1947, Chapman Papers, University of Washington Libraries.

91  See, *inter alia*, Edward Wenk, Jr., *The Politics of the Oceans* (Seattle and London, 1972); Shor, *Scripps Institution*; and annual reports of SIO, 1950s-present.

92  See Wenk, *Politics of Oceans, passim*; U.S. Commission on Marine Science, Engineering and Resources, *Our Nation and the Sea: A Plan for National Action* (Washington, 1969), 1-48; Scheiber, 'Pacific Ocean resources, science and law of the sea,' 406-27. The physical oceanographer Carl Eckart, who followed Sverdrup, albeit for only a short time, as director of SIO, reflected in June 1948 on the direction of the Institution's ecosystem research effort: 'The individual scientist, working in seclusion,' he declared, 'is apparently a thing of the past'. Nor was he certain that this was 'going to be good for science'. (Eckart to Walford, 28 June 1948, SIO Directors Files: Marine Life Research, SIO Archives.)

93  Memorandum from Oscar E. Sette and others, 'The need for oceanographic studies for Pacific coast fisheries', 1947, in SIO Directors Files, SIO Archives (asserting, with entire accuracy, the possibilities for new techniques); John Isaacs and Columbus Iselin, eds, *Oceanographic Instrumentation* (Washington, 1952)

(includes discussion of mid-depth trawl, developed by an SIO engineer and a federal agency biologist); Scheiber, 'California marine research', 63 *et passim*.

94 See contributions by John Knauss, Michael Mullin, and others, in CalCOFI Reports, 31 (1990), (Symposium panel on global change, and also individual articles).

# CORNISH AND SCILLONIAN MARINE STUDIES, PAST AND PRESENT

## Stella Maris Turk

## ABSTRACT

As human pressures have increased, there is overwhelming evidence that, worldwide, wildlife habitats and wildlife species have decreased. The effect of the wreck of the *Torrey Canyon* in 1967, resulting in the then largest oil spill, is discussed and an historical outline is given of recording before and since the oil spill, including the role of the Cornish Biological Records Unit and its computerisation project. Reasons for Cornwall's rich diversity of marine life are summarised, and various changes - natural and due to human intervention - in the marine environment are considered. Deterioration of Cornish shore life is discussed under the multiple factors that are believed to contribute to this. The Helford River is used as an example of past recording showing present changes, and protective action being taken.

## INTRODUCTION

### Global changes

Worldwide, the more sensitive species are becoming extinct faster than they could ever be replaced. Land, freshwater and marine environments are becoming dominated by adaptable, pollution-proof species which are often those that could be termed 'slum dwellers' because of their ability to live in degraded conditions. Humans have compounded the problem by deliberate or accidental introduction of animals and plants, many of which threaten to replace native species, sometimes by producing vigorous hybrids, like Cord-grass (*Spartina anglica*).

## Changes, gradual and cataclysmic

Competing for space and resources are everyday topics in our local and national debates concerning development, available funds and concern for the environment. Industrial activities of one sort or another in Cornwall have been responsible for much despoilation of land, sea and shore, and whilst tourism is not the most dramatic in this respect, it is now probably the biggest of all industries, and it can cause special problems since it generally caters for living conditions of a high standard in areas which are of great natural beauty and diversity: in Cornwall these sites are often by the open coast or within those productive fingers of the sea (rias) that fill the deep sunken valleys of south Cornwall - e.g. the Fal Estuary and the Helford River. The effects on the coast from most causes are subtle but insidious, but others are sudden and have immediate impact: the diversity of a shore can be dramatically reduced by the catastrophe of an oil spill.

## 1967: wreck of the *Torrey Canyon*: the first major oil spill

Britain was first confronted with the problems of a major oil spill when the *Torrey Canyon* was wrecked on the Seven Stones reef in 1967. Scientists at the Laboratory of the Marine Biological Association of the U.K.(MBA) immediately started assessing the damage to Cornish wildlife caused by the oil and dispersants (Smith, 1968). Detailed monitoring of the recovery rate of shore barnacles, was instigated by Dr Alan Southward and Dr Eve Southward. Although staff at the MBA Laboratory at Plymouth have been concerned with marine life of the surrounding area since it was founded in 1884, there were few shore surveys further west than Looe, since visits to Cornwall were very time-consuming until the road bridge over the Tamar was opened in October 1961. However, one of the sites affected by the oil, was Newtrain Reef, Trevone (on the north Cornish coast) where Dr Douglas Wilson had recorded and photographed a particular gully over many years; and another was Marazion where Mr Peter Corbin had recorded fauna from the *Zostera* bed just before the disaster. The absence of even simple surveys and lists from the other areas affected, was a serious drawback, and was to set the scene for a new evaluation of, and care for, the marine environment.

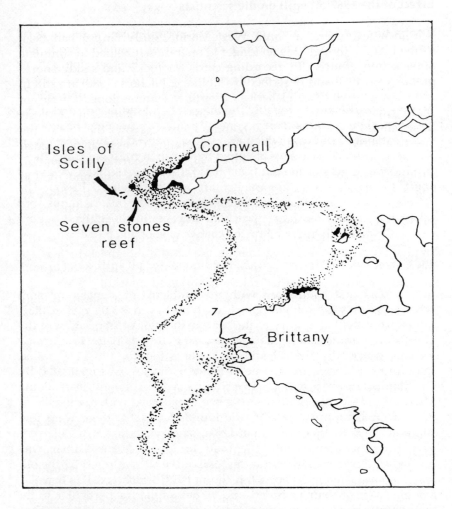

Figure 1. *Torrey Canyon* oil spillage. (Redrawn from fig. in Smith, J.E., 1968). March 1967 saw the world's largest oil spill which shocked the scientific world. The repercussions continue.

**Effect of the 1967 oil spill on the scientists**

Within a few years, a South West Marine Study Group had been formed, and although it is no longer in operation it helped to stimulate other action groups. Its recording cards for rocky and sandy shores prepared in 1978 and described by Holme & Nichols (1980) are still in use. The Cornwall Sea Fisheries Committee commissioned a series of shore transects to provide a basis for future monitoring (Atkins Research & Development, 1976), and a few years later the first of the intertidal surveys, commissioned by the Nature Conservancy Council was carried out (Powell *et al.*, 1978). Both the Marine Conservation Society (formerly the Underwater Conservation Society) and the Oil Pollution Research Unit (OPRU) were under contract to the NCC to produce surveys of Cornish and Scillonian shores and shallow seas, particularly bays, harbours, rias and estuaries. Now the Marine Nature Conservation Review (MNCR), initiated and funded by English Nature, is assessing marine life all around the British coast and will publish its findings in a series of volumes. The MNCR will take into account all earlier relevant surveys, some of which were directly related to the move by the NCC to provide extra protection to the marine environment by creating statutory Marine Nature Reserves, in this instance around the Isles of Scilly: the move was abortive and the Isles have recently been designated a voluntary-style Marine Park.

**Effect of the 1967 oil spill on local naturalists**

Throughout this period and as a direct result of the lack of basic available information, when the *Torrey Canyon* disaster demonstrated the need, the author was visiting many Cornish shores, making general ecological assessments and compiling species lists which incorporated, where appropriate, observations by earlier naturalists from the particular site, e.g. Turk, S.M. (1977). This was often undertaken as part of the study-work of classes organised by the University of Exeter, the LEA or WEA. Later it involved members of the Cornish Biological Records Unit, sometimes in conjunction with the Cornwall Trust for Nature Conservation or one of the three Cornish natural history study groups (Lizard Field Studies Club, Camborne-Redruth Natural History Society or Caradon Field and Natural History Club). More recently Heritage Coast outings have been involved. The sum total of this work, housed at the CBRU, has been extensively used in the compilation of all the NCC surveys, as has the Unit's comprehensive bibliography of works on Cornish marine biology. In 1987, the Cornwall Trust for Nature Conservation arranged for an MSC-funded post, under the Manpower Services Commission's Community Programme, for the categorisation of Cornish marine source material held by the CBRU; John Polglase was based at the Unit to prepare this index.

## ROLE OF THE CORNISH BIOLOGICAL RECORDS UNIT

### Establishment of the Unit, and its aims

The Unit was established in 1972, as one of the indexes of the Institute of Cornish Studies (directed by Professor A.C.Thomas until October 1991, when Dr Philip Payton became Director, and jointly funded and managed by Cornwall County Council and the University of Exeter) under the immediate direction of Dr Frank Turk. It was to occupy a quarter of his teaching time as an Extra Mural tutor and Reader in Natural History Studies, University of Exeter. He planned it on a wide basis, to include records of all species of plants and animals that have lived or do live in Cornwall and the Isles of Scilly, their shores and adjacent seas. Thus one-celled plants and animals as well vertebrates, modern, subfossil and fossil, are recorded, together with parasites, commensals and epiphytes - truly everything, and now approaching 21,000 species. Sources for records must be equally catholic, and owing to the popularity of Cornwall, the literature is spread throughout British national journals, as well as some Continental ones. In addition to the records of species, copies of all known marine surveys of Cornish sites are kept in the archives. Relevant journals (including a complete set of the *Journal of the Marine Biological Association*) and offprints of papers on Cornish marine and non-marine life are also in the data bank. An archive of photographs and relevant newspaper cuttings has also been formed. Having helped with the work of the Unit from its inception the author became wholly responsible for its direction in 1985 when Dr Turk retired.

### Computerisation of the records — ERICA

By the mid-1980s, over a million biological records were on the manual index, with some 30,000 published and unpublished references in the bibliography, all achieved by voluntary and part-time help. This massive collection of records is recognized as being the largest and most comprehensive regional data bank in Europe. Dr Colin French, the only full-time member of the Unit, was appointed to create a programme to computerise the records without losing any of the varied data on the manual index. His ERICA (Environmental Recording In Cornwall Automated) is necessarily complex although relatively easy to use for the operators, most of whom are from Cornwall County Council's Employment Training Agency, being placed with us to learn or enhance computer skills. Currently 400,000 records of 17,000 plants and animals are on our mini-computer, accessed at Exeter by direct line. This has been made possible by generous sponsorship from Prime Computer UK and British Telecom. An account of the recent progress of the ERICA project has been written by French (1991).

## Sources of marine biological records

The preponderance of marine surveys and the vast number of marine records in the CBRU reflect the current and past interest in Cornish shores and shallow seas. Cornwall became the 'Mecca' of marine biologists in the eighteenth and nineteenth centuries. Those domiciled in Cornwall included the Rev. William Borlase, Dr Jonathan Couch and Dr W.P.Cocks whilst many more like Prof. Edward Forbes, Joseph Alder and the Rev. Alfred Cooke visited Cornwall. In the first few years of the twentieth century, Dr James Clark of the Truro Technical Colleges, did an enormous amount of general biological recording - a great part of it marine - with his students, and he published the main results of this work in the *Victoria County History of Cornwall* (Clark, 1906). He also included a summary of most earlier records from Borlase (1758), Couch (1841-4) and Cocks (1849). From 1910 to the early 1950s comparatively little shore-recording work was carried out in Cornwall and the Isles of Scilly, although the Victorian interest in molluscs was continued by A.P.Gardiner and T.G.W.Fowler. It was the mid 1950s when the 'diver biologists', often in Sub-Aqua Clubs, became concerned in recording with an increasing amount of professional involvement. At the same time, Professor L.A. Harvey started work on the Scillonian marine fauna and summarised the ecology of the Isles of Scilly (Harvey 1969a) in his introduction to a series on the marine fauna of the Islands which has appeared in the *Journal of Natural History*. In 1957 the much-consulted 3rd edition of the *Plymouth Marine Fauna* was published (Marine Biological Association, 1957). An historical survey of marine algae recorded from the Lizard peninsula (Price et al., 1979; 1980) is another important study. Records resulting from all this work, spanning two centuries and more, have been centralized at the CBRU, and are rapidly being transferred to computer.

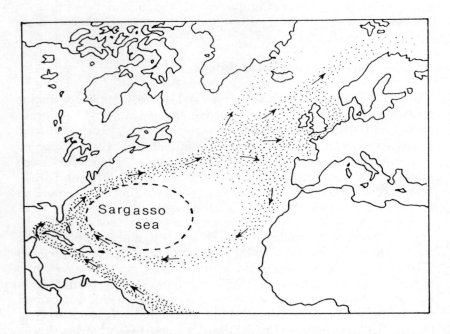

Figure 2. North Atlantic Drift. The Drift which spreads northward from the Gulf, brings with it many casual warm-water species as well as enabling others to live on the shores and in the shallow seas of western Britain, especially in the far south-west.

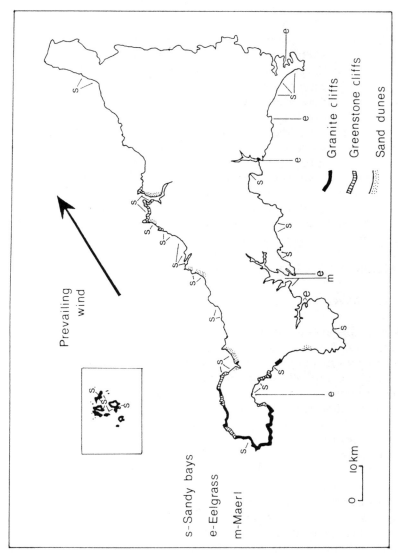

Figure 3. Cornwall's long coastline. The indented coastline of Cornwall provides varying degrees of shelter and substrate for a wealth of marine life.

## THE IMPORTANCE OF CORNISH MARINE LIFE

### Reasons for Cornwall's rich diversity of marine life

Cornwall, quantitatively and qualitatively, has a wealth of marine habitats, due, in part, to its long coast line (longer than any English county) and its variety of substrates. Add to this its geographical position and the fact that its shores are bathed by the North Atlantic Drift, and one would expect to find warm-water species, both casual and resident. Thus there are more turtles recorded off the Cornish coast than any other part of the British Isles (Penhallurick, 1990, 1991), and there are many 'Mediterranean' organisms living on and off shore. Moreover some marine animals that normally live off shore, can be found between tide-marks on the Cornish and Scillonian coasts - believed to be a reflection of the comparatively frost-free conditions. Many 'northern' species also live this far south, Cornwall being noted for its mix of 'northern' and 'southern' species - a general condition reflected in the plankton. Background information on the marine life of Cornwall and the Scillies is given by Turk, S.M. (1971 & 1983).

## CHANGES IN THE MARINE ENVIRONMENT

### Natural changes

We know all too little of these. There seem to be cyclic phenomena perhaps related to sun-spot activity and broadly relatable to climatic shifts, involving changes of temperature, water-movements, chemical constituents, competition and natural fluctuations (Southward, 1980). A species may extend its range consequent on 'population explosion' movements that may follow exhaustion of food supplies: such phenomena are known in octopuses (Garstang, 1900; Rees & Lumby, 1954). Tolerance of previously limiting conditions may be achieved, e.g. the American Hard-shelled Clam, *Mercenaria mercenaria*, that needed 20°C. to breed in the USA, will now breed in Southampton Water at 16°C. (Al-Sayed, 1988). As on land and in freshwater, there are seasonal changes in the sea and on the shore. Obvious shore differences are shown by annual and perennial species of seaweeds reaching their peak at different times of year. Algal blooms will alter the appearance of a shore, as will changes in the levels of sand and the movements of freshwater streams.

## Short-term effects

The fact that the lowest spring tides are at midnight and midday in Cornwall and the Isles of Scilly, means that the heat of the summer sun can scorch, whilst the cold of a winter's night can damage the plants and animals that are exposed to the elements as the tide falls. Even the well-adapted organisms of the upper shore may die; more so the vulnerable fauna and flora of the lower shore less adapted to exposure. The effects of extreme cold were well documented during the winter of 1962/63 by Crisp and Southward (1964). Hawthorne (1965) showed how the eastern limit of the Thick Top-shell (*Monodonta lineata*) is governed by temperature: it increases eastwards over a period of fine summers and comparatively frost-free winters, only to suffer heavy mortality during a winter of severe cold.

## Long term effects? Global warming?

Cornish waters would seem to be an ideal environment in which to monitor global warming because there are so many warm-water and cold-water species living side by side; but when considering historical records, past data are often imprecise regarding identification and/or locality. Some species, however, are distinctive enough always to attract public attention and often carry sufficiently accurate written records even in the newspapers. Such are the great Leathery Turtles (*Dermochelys coriacea*), sightings and strandings of which span a couple of centuries. Penhallurick (1990; 1991) shows that from 1756 there were never more than three reports in any one year - and often there were none - until 1988 when there were 16, a figure repeated in 1990. They feed on jellyfish, particularly the large 'Rootmouth', *Rhizostoma octopus*, examples of which can reach up to 90 cm in diameter and which, in turn, feed on plankton. These jellyfish have been present in vast numbers in recent years, weighing down trawls with their massive weight, some even remaining close to our coasts through the colder months of the year.

Much media attention has been given to the many stranded dolphins on the Cornish coast during the last couple of years. Interpretation is not easy, but it could be that whilst the overall number of dolphins is believed to be decreasing, more of them have been moving northwards, following their food of cephalopods or fishes. Whether this warming is short-term or long-term, only time will tell; but one thing is certain: just as the turtles run the risk of being entangled in crab-pot ropes, so the dolphins can become trapped in nets, explaining a proportion of the strandings.

Swaby & Potts (1991) describe how an increase of just half a degree centigrade in the Western Approaches has been sufficient 'to force northern species north and allow exotic southern species to extend their range to our southern shores'. Thus several warm-water fishes have increased in number, including the Trigger-fish (*Balistes carolinensis*), the Sea-horse (*Hippocampus ramulosus*) and the Puffer-fish (*Lagocephalus lagocephalus*). In 1990, a rare garfish (*Belone svetovidovi*) was caught in Mount's Bay, a first record in British waters.

Organisms which attract attention because of their unusual nature, are those oceanic drift species, By-the-Wind Sailors (*Velella velella*), Violet Sea-snails (*Janthina spp.*) and Buoy Barnacles (*Lepas fascicularis*). These are now found most years, and *Velella* may reach astronomical numbers, as in 1981 (Turk, 1982). They are never as common on the more sheltered south shores of Cornwall, but it seems surprising that between 1843 and 1878 when Dr W.P.Cocks was scrupulously recording all marine phenomena in the Falmouth area, he recorded *Velella* only twice. Species of *Janthina* are also abundant on occasion, as reported by Wilson & Wilson (1956) and Wilson (1958). The Portuguese Man-of-War (*Physalia physalia*) is another species which attracts much attention, although many reports refer to jellyfish: a large invasion was reported by Wilson (1947). The larger drift seeds are also noticeable, and again Cornwall has most of the strandings (Nelson, 1986). A more dramatic effect of the North Atlantic drift was demonstrated in December 1986 when a 35-foot power boat was found drifting upside down off St Ives. It was covered with a mass of goose barnacles (removed by a commercial firm to sell as a delicacy) as well as shells and other marine species native to the West Indies. There was speculation that it may have been used for smuggling dope or illegal immigrants since it had ribs from the fibre-glass hull cut away to make more space and it had three 120 HP engines. A published note (Turk, 1988) brought a response from the editor of *New York Shell Club Notes* which reprinted the information, confirming that 'these powerful speedboats pick up contraband well off shore and then come into a quiet cove or residential marina'.

**Changes due to introductions**

Some changes are unquestioned, and we even know their precise history. Such are certain introductions, one of which is the the American Slipper Limpet (*Crepidula fornicata*), which competes with oysters for food, and which was imported with the American Blue-point Oyster (*Crassistrea virginica*) about 1890. The first Cornish record refers to shells found in beach rubbish in the Helford River in 1935: Gardiner (1935) writes, 'St G.Byne found...five *Crepidula fornicata* (Linne) in beach rubbish at Helford. These were probably brought in with oysters introduced into the oyster farm at Port Navis (sic). Great care

Figure 4. Leathery Turtle. These huge reptiles can attain 2.60m. in length and their occurrence always attracts attention. In 1988 and again in 1990 there were 16 Cornish records, compared with a maximum of three for any previous year back to 1756.

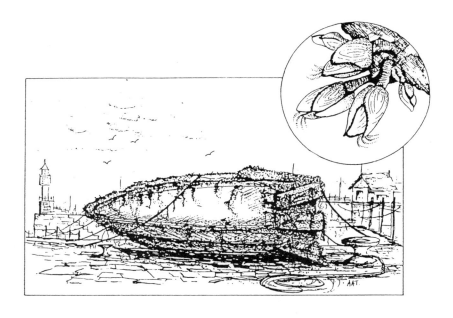

Figure 5. American power boat, redrawn from a photograph. This fibre-glass boat was evidently used for smuggling, and reached Cornwall, in December 1986, via the North American Drift. It was covered with goose barnacles as well as many typical species from the West Indies.

should be taken to guard against the introduction of this species into the beds'. *Crepidula* was not recorded as breeding in the Helford until 1966 (T.E.Thompson, Bristol University card index) but it has been known in the Fal since 1947 (McMillan, 1956). The Australian Barnacle (*Elminius modestus*) arrived on ships in Chichester Harbour in 1947 and is now abundant in Cornish sheltered waters, whilst the Japanese Tunicate (*Styela clava*) may be abundant on the south British coast, the first Cornish records were in 1972 and Smith (1991) reports it is now found inter-tidally as far north as Loch Ryan. The latest of several introductions that are now naturalised is Japweed (*Sargassum muticum*), a relative of Gulfweed: it was first found on the British coast in 1973 and reached Cornwall in the late 1970s (Boalch, 1986). It has an effect on our native flora and fauna because of its rapid growth and because it does not attract epiphytes to the same degree as the native seaweeds it can replace. Gubbay (1988) lists 11 introduced species of Algae, 8 of which - from the Pacific and Atlantic coasts of America and Australia as well as Japan - grow on Cornish shores; and 13 animal species, 6 of which are 'Cornish'. All these are accidental introductions apart from the Blue-point Oyster (*Crassostrea virginica*) which has been imported together with the Pacific or Japanese Oyster (*Crassostrea gigas*). The Portuguese Oyster (*C. angulata*) is now accepted as a synonym (or, at most, a subspecies) of the Pacific Oyster following research which was summarised by Menzel (1974): his view that that 'Ports' might have been taken to Japan in the sixteenth century, is discounted as they are known from ancient Japanese kitchen middens (Kidder, 1959). Could they have been taken from Japan to Portugal?

## DETERIORATION OF CORNISH SHORE LIFE

### Effect of holiday makers and educational groups

If the geographical, geological and physical features of Cornwall lead us to expect a rich flora and fauna, we would anticipate that the pressures that are now present on our shores, would be accompanied by deterioration. Harvey (1969b) wrote, 'Tourism, although a mild price to pay for freedom from urbanisation, nevertheless carries its penalties in the thoughtless erosion of natural areas by holiday makers, a phenomenon which is particularly apparent on some shores where the constant turning and treading of boulders and stones has seriously diminished the flora and fauna'. Whilst leading an excursion to Gyllyngvase Beach, Falmouth, Clark (1907) described it 'as one of the finest beaches in Great Britain for shore collecting, with Falmouth Bay one of the richest dredging grounds for the naturalist in the British Isles'. In 1975, when Pitchford (1977) visited nearby Swanpool Beach it was 'to find holiday-makers from the nearby holiday camp had turned over almost all of the boulders and stones, thus destroying almost all the creatures that should have been found there.... These rocks with the adjacent Gyllyngvase Beach would have been a naturalist's paradise years ago when largely unspoilt by extensive human intrusion'. It has to be remembered that it is not only visitors to Cornwall who enjoy its shores and shallow waters: those domiciled here also seek recreation in, on and under this precious resource.

In the past, educational groups have been more concerned with finding as much as possible, and even removing large numbers of organisms for examination. Now with a greater awareness of the need to care for the environment with propaganda from such prestigious organisations as the Marine Conservation Society, one hopes that more care for the shore is emerging, allowing study to become compatible with conservation.

## Pollution and human disturbance

### Industrial pollution

The danger of chemical pollution even so far from industrial centres, was emphasized by the destructive properties of TBT (tributyltin) paints, now banned on all craft below 12 m. Its use was greatest in rias, estuaries and harbours where oysters were farmed, and its effect on the Native Oyster was devastating. It also caused the females of the Purple Dog-whelk (*Nucella lapillus*) to take on male characteristics to the extent that they became sterile (Gibbs & Bryan, 1986). Cornwall is no stranger to pollution by heavy metals: for instance, mine drainage has caused the Restronguet Creek to become one of the most heavily polluted estuaries in Great Britain (Holliday & Bell, 1981) whilst Mevagissey Bay received 17 million tons of china clay waste from the mid eighteenth century until the 1970s when discharge into the sea was discontinued (Holme & Probert, 1978).

### Slurry, sewage and silage

Small quantities of organic-rich effluents add valuable nutrients to the sea, but when there are large concentrations, the balance can be upset, especially in bays, harbours, estuaries and rias. Unhealthy enrichment can be the immediate result. Mark Hannam, South West Water's principal oceanographer, as reported in *The Western Morning News* for 28 August 1991, stated that 'The problem of waste water management is most acute in the South West region because there is a natural dependence on the area's extensive coastal waters'. Even on the wild exposed north coast of Cornwall, the additional sewage effluent that occurs during the holiday season (when the population doubles) has its effect: pollution from increased sewage outfall can be measured by the decrease in diatom diversity (Hendey, 1977). It is now accepted that certain harmful bacteria and viruses can remain active for a considerable time in the sea.

### Shore-collecting of bait, shellfish and seaweed

As angling has become more popular, bait-digging has increased. Anglers have a strict code of behaviour but, nevertheless there can be considerable disturbance of an area, and many diggers still do not back-fill the holes.

Searching for shellfish generally involves moving weed, turning stones or raking, thus affecting many other creatures directly and indirectly by the destruction of habitats. Molluscs have been an important component of human diet for millenia. Middens in the Isles of Scilly have a preponderance of very small limpet shells, suggesting that, unless the ancient inhabitants deliberately collected small specimens, the populations were heavily predated. Certainly now very large limpets are to be found on Scillonian shores. Townsend (1967) and Ashbee (1968) discuss the role of limpets as a main article of diet and Turk, F.A., 1984) believes that they were also used as ground bait. Now few limpets are collected for human consumption or bait in the Scillies and mainland Cornwall although for centuries, and until a few decades ago, they were an important part of the Cornish custom of 'trigging' on Good Friday. Trigging is collecting such shellfish as cockles, mussels and winkles as well as limpets. The Good Friday practice continues, cockles now being the main species, with suspected deleterious effects on the environment. Commercial collecting is mainly concerned with winkles, collected by the ton for the French market.

At the end of the nineteenth century, the demand for seaweed as fertiliser for potato crops, was so great that Ralfs (1880-5) wrote that '...when under the name of Oreweed it is procured by stripping bare, not only the rocks exposed at ebbtide but, by means of a long-handled basket-scythe, those in deep water as well, I greatly fear by thus depriving the young fish of shelter and good feeding ground their annual migrations may be directed to more protected coasts and thereby materially affect the harvest of the sea'. Now not even the cast-up weed is collected, except in very small quantities, and certainly none is cut. However, there is considerable demand for fertilizer in the form of the nodules of dead calcified seaweed known as maerl, dredged in the Fal Estuary. *Living* beds once filled the mouth of the

*Plate 1.* Gillan Creek. This small creek is part of the Helford River complex showing the typical marine conditions deep in a tree-lined valley.

Estuary and much of the Bay, but there is now only a remnant left on St Mawes Bank, part of the Roseland Marine Conservation Area. Both living and dead deposits are of great marine biological importance as they harbour such a diversity of plants and animals. Unlike Wales and North Devon, there has never been more than a minimal demand for seaweed food in Cornwall.

## HELFORD RIVER: A CASE HISTORY FOR ASSESSING CHANGE

Unfortunately, except for those plants and animals that attract the attention of the scientist or the media, the written records of plants and animals of the nineteenth century and earlier decades of the present century rarely carry the detail of exactly where and when found, their numbers and associated species. Nevertheless it can be claimed that records kept by the CBRU presented sufficient evidence of change to instigate protective measures for the Helford River.

### Helford, famous for its marine life

Helford River is a drowned river valley, or ria, formed by rising sea-levels after the last glaciation. It is virtually fully marine, as well as relatively sheltered, and its outstanding marine biological significance was recognized by the Nature Conservancy Council as being of International Importance (Powell et al., 1978; Bishop & Holme, 1980). Marine records date from the late 1840s, but due to its remoteness, it received less attention than Mount's Bay and Falmouth Bay. Gardiner (1927) wrote of the richness of the area, but the first systematic surveys were not carried out until 1949 (Spooner & Holme, 1986). Holme & Turk (1986) list all records and references up to 1910 and Tompsett & Turk, S.M.(in prep.) are compiling a check-list of all fauna, in date phases. Throughout the 1960s and into the early 1970s, Bristol University students visited Cornwall annually, recording the rich marine fauna of the Falmouth and Helford areas (Bristol University Card Index). The author accompanied them on more than one occasion, and all the records were in due course given to the CBRU. In 1972 as joint tutor in a University Field Course with Roger Burrows, a visit was paid to Helford Passage and detailed notes were made of the wealth of animal life (Burrows & Turk, S.M.1972).

## Warning signs

In the mid-1970s Roger Burrows and Dr Paul Chanin reported that the beds of Eel-grass (*Zostera marina*) at Helford Passage were becoming eroded and, inevitably, the associated life had decreased. Soon afterwards, the late Dr T.E.Thompson of Bristol University, asked where he could go with his students to find a good deposition shore, as Helford shores had become so degraded that he feared that they would become as azoic as those of northern France where collecting for food was so intense. Dr Thompson had been bringing students to the Helford area for over a decade, and had reason to know the area well. When the author (Turk, S.M., 1984) visited Helford Passage after a lapse of 12 years, deterioration of surface life was obvious. Comparison with earlier reports (Holme, 1987; Holme & Bishop, 1978; Turk, S.M., 1972, 1976, 1977) on other deposition shores - at Treath, Penarvon and in Gillan Creek - showed a similar degradation, with a great increase of the Sand-mason Worm (*Lanice conchilega*) and decrease of the formerly fine Zostera beds. By 1987, all the inter-tidal *Zostera* had disappeared, and Prof. C.den Hartog (den Hartog, 1989), to whom samples were sent in 1989, believed this was due to 'wasting disease' one of the causative agents of which is a species of *Labyrinthula*, a micro-organism probably related to the slime-moulds. This wasting disease was widespread in the 1930s (Wilson, 1949) although Turk, S.M.(1986) showed that it remained (or became re-established) in most of its earlier sites although probably with less vigorous growth. *Labyrinthula* is believed to be endemic, only reaching dangerous epidemic proportions if the plants are under stress. Human activities alone are known to cause serious decline in seagrass beds all over the world (Short et al., 1988) but insolation, frost and excess rainfall can all cause stress.

## Establishment of a Voluntary Marine Conservation Area

In 1983 an approach was made to Cornwall County Council with the suggestion that the Helford River should receive some measure of protection and the same year Holme (1983) wrote of the desirability of such a move. The County Council convened a series of meetings of interested groups and individuals, and in 1987 the Helford River was designated a Voluntary Marine Conservation Area, following the findings of a survey (Covey & Hocking, 1987) which presented evidence of deterioration of the silt/sand/gravel areas, based on earlier reports and records.

## Monitoring

The aim of the VMCA is to achieve by voluntary means the harmonious use of the river and to monitor the quality of the marine environment.

Three monitoring reports have now been produced (Hocking, 1989; Turk; 1990; Tompsett, 1991). These involve repeat recording of fixed transects with photographs of selected quadrats.

## Strategic Guidelines

The voluntary nature of the scheme precludes management plans, so a series of Guidelines have been prepared suggesting ways to protect the River whilst reconciling the many interests.

## Hopes for the future

Experiments are underway to diversify the shore by adding slaty stone such as was on much of the shore when the silt was 'fixed' by *Zostera* roots. The understandable suspicion which met the attempts to establish the VMCA have been allayed, and there is encouraging collaboration by users of the River.

The work has received sponsorship from many bodies, local and national. The initial work was funded by the World Wide Fund for Nature (WWF), in part with Heinz Guardians of the Countryside, whilst the first monitoring Report was contracted by the Nature Conservancy Council and WWF. WWF has continued to grant aid the project and now (1991/92) it has arranged commercial sponsorship from the National Westminster Bank.

The trauma of the massive oil spill from the *Torrey Canyon* alerted scientists to the vulnerability of shores and shallow seas and to the lack of even simple surveys. This acted as a catalyst to build up such information nationally as well as regionally. The Cornish Biological Records Unit, formed in 1972, has centralised much information on the marine life of Cornwall, spanning the past couple of centuries. Coastal areas, especially the drowned valleys and sheltered bays of Cornwall, are vulnerable to the more insidious pressures that arise from human activities, which include the introduction of species. Changes due to human intervention are not always easily separated from natural fluctuations of species and populations, and the importance is emphasized of those species which attract the media and which therefore may provide some statistical basis of change. The Helford River is used as a model of changes noted due to recording and remedial steps being taken.

## ACKNOWLEDGEMENTS

Varied help from Pamela Tompsett is gratefully acknowledged. I am also indebted to Daniel Flunder for the photograph of Gillan Creek (Plate 1) and to Andrew Tompsett for all the other illustrations.

Sea-horse. This warm-water fish is sometimes found in the sheltered bays and rias of the south-west, living and possibly breeding in eel-grass beds.

## REFERENCES

1 Al-Sayed, H.A.Y., 1988. 'Population studies of a commercially fished bivalve, *Mercenaria mercenaria* (L.), in Southampton Water', Ph.D thesis, University of Southampton.

2 Ashbee, P., 1968. 'Excavations at Halangy Down, St Mary's, Isles of Scilly', *Cornish Archaeology*, 7, 24-32.

3 Atkins Research & Development, 1974. *The Pollution of Cornish Coastal Waters*. A Report presented to Cornwall Sea Fisheries Committee, A.R. & D.Epsom.

4 Atkins Research & Development, 1976. *The Pollution of Cornish Coastal Waters*. *Final Report presented to Cornwall Sea Fisheries Committee*, A.R. & D.Epsom.

5 Bishop, G.M. & Holme, N.A., 1980. *Survey of the Littoral Coast of Great Britain. Final Report - Part 1 The sediment shores - an assessment of their conservation value*. Report to the Nature Conservancy Council. 77 pp. + 20 pp. of maps.

6 Bristol University Zoology Dept Card Index. *Records of Cornish Marine Fauna*, 1983 ff.

7 Boalch, G.T., 1986 (for 1983 & 1984). 'The story of Japweed in Cornwall', *The Lizard*, 7 (3 & 4), 21.

8 Borlase, W., 1758. *The Natural History of Cornwall* (London).

9 Burrows, R. & Turk, S.M., 1972. 'Helford Passage'. Unpublished typescript.

10 Clark, J., 1906. 'Marine zoology', In Page, W., ed. *The Victoria History of the County of Cornwall* (London).

11 Clark, J., 1907. 'Excursion to Gyllyngvase Beach', *Report Royal Cornwall Polytechnic Society (RRCPS, 75, 47-51)*.

12 Cocks, W.P., 1849. 'Contribution to the fauna of Falmouth', *RRCPS*, 17, 38-102. A series of papers on the fauna of Falmouth, mainly marine, followed for most years up to 1877.

13 Couch, J., 1838. *A Cornish Fauna. Pt 1. Vertebrate, crustacean and a portion of the radiate animals* (Royal Institution of Cornwall (RIC) Truro).

14 Couch, J, 1841. *A Cornish Fauna. Pt 2. The testaceous mollusks* (RIC, Truro).

15 Couch, R, 1844. *A Cornish Fauna. Pt 3. The zoophytes and calcareous corallines* (RIC, Truro).

16 Covey, R. & Hocking, S.,1987. *Helford River Survey Report. Report to the Helford River Steering Group* (Cornish Biological Records Unit, iv + 121 pp.).

17 Crisp, D.J. & Southward, A.J., 1964. 'The effects of the severe winter of 1962-1963 on marine life in Britain', *J.Anim.Ecol.*, 33, 165-210.

18  den Hartog, C., 1989. 'Wasting disease in *Zostera marina'*, *Plant Press*, 7, 4.

19  French, C., 1991. 'ERICA, facts, figures and developments', *Zoo.Cornwall & I.of S.*, 1, 2-3.

20  Gardiner, A., 1927. 'Ecology notes on the *Zostera* beds in the Helford River, Cornwall'. *J.Conch.,Lond.* 18: 147-8.

21  Gardiner, A.P., 1935. 'Marine Recorder's Report', *J.Conch., Lond.*, 20, 189-90.

22  Garstang, W., 1900. 'The plague of octopus on the south coast and its effect on the crab and lobster fisheries', *Journal of the Marine Biological Association of the U.K.*, (*JMBA*), 6, 260.

23  Gibbs, P.E. & Bryan, G.W., 1986. 'The decline of the gastropod *Nucella lapillus* around South-West England: evidence for the effect of tributyltin from anti-fouling paints', *JMBA*, 66, 611-40.

24  Gubbay, S., 1988. *Coastal Directory for Marine Conservation*, (Marine Conservation Society).

25  Harvey, L.A., 1969a. 'The marine fauna and flora of the Isles of Scilly. The Islands and their ecology', *J.Nat.Hist.*, 3, 3-18.

26  Harvey, L.A., 1969b. 'Zoology', *Exeter and its Region* (British Ass./University of Exeter).

27  Harvey, R., Knight, S.J.T., Powell, H.T. & Bartrop, J., 1980. *Survey of the littoral coast of Great Britain. Final Report - Part II The rocky shores - an assessment of their conservation value* (Report to the Nature Conservancy Council) 55 pp.

28  Hawthorne, J.B., 1965. 'The eastern limit of distribution of *Monodonta lineata* (da Costa) in the English Channel', *J.Conch., Lond.*, 25, 348-52.

29  Hendey, N.I.,1977. 'Diatom communities and its use in assessing the degree of pollution insult on parts of the north Cornish coast of Cornwall', *Nova Hedwigia*, 54, 355-78.

30  Hocking, S., 1989. *Helford River Survey. Monitoring Report No.1. A Report to the Nature Conservancy Council with additional funding from WWF* (Cornish Biological Records Unit).

31  Holliday, R.J. & Bell, R.M., 1981. *The Ecology of Restronguet Creek and Fal Estuary* (Environmental Advisory Unit, University of Liverpool/Billiton Minerals).

32  Holme, N.A., 1983. 'A marine nature Reserve on the Lizard?', *The Lizard*, 6 (4), 15-17.

33  Holme, N.A., 1987. 'Gillan Harbour: changes in fauna and flora'. Unpublished report, including survey notes from 1951-1986.

34  Holme, N.A. & Bishop, G., 1978. Penarvon Cove. Unpublished field data.

35  Holme, N.A. & Nichols, D., 1980. Habitat survey cards for the shores of the British Isles. (Field Studies Occasional Publications) No.2, 1-16.

36    Holme, N.A. & Probert, P.K., 1978. 'Disposal of solid waste in the marine environment with particular reference to the china clay industry', in Goodman, G.T. & Chadwick, M.J., eds., *Environmental Management of Mineral Wastes* (The Netherlands: Sijthoff & Noordhoff).

37    Holme, N.A. & Turk, S.M., 1986. 'Studies on the marine life of the Helford River: fauna records up to 1910', *Cornish Biological Records*, 9, 1-26.

38    Holme, N.A. & Turk, S.M., 1989. 'Helford Voluntary Marine Conservation Area: its history and future', *The Lizard*, 3rd Series, 1 (pts 3 & 4), 24-7.

39    Kidder, J.E., 1959. *Ancient Peoples and Places* (London, Thames & Hudson).

40    McMillan, N.F., 1956. 'Marine aliens in Cornwall', *J.Conch., Lond.*, 24, 80.

41    Menzel, R.W., 1974. 'Portuguese and Japanese oysters are the same species', *J. Fish. Res. Board Canada*, 31, 453-6.

42    Miller, P.J. & El-Tawil, M.Y., 1974. 'A multidisciplinary approach to a new species of *Gobius* (*Teleostei gobiidea*) from southern England', *J.Zoology*, 174, 539-74.

43    Nelson, E.C., 1986. Catalogue of European drift seeds. Computerised list.

44    Penhallurick, R.D., 1990. *Turtles off Cornwall, the Isles of Scilly and Devonshire* (Truro: Dyllansow Pengwella).

45    Penhallurick, R.D., 1991. 'Turtle occurrences off Cornwall and Scilly in 1990', *Zoo. Cornwall and I.of Scilly*, 1, 6-10.

46    Pitchford, G.H., 1977. 'Some Cornish beaches', *Conch.Newsletter*, 63, 49-53.

47    Marine Biological Association, 1957. *Plymouth Marine Fauna* (3rd edn.).

48    Powell, H.T., Holme, N.A., Knight, S.J.T. & Harvey, R., 1978. *Survey of the Littoral shores of Great Britain. 2. Report on the shores of Devon and Cornwall. A Report to the Nature Conservancy Council from the Marine Biological Association/Scottish Marine Biological Association*, iv + 126 pp.

49    Price, J.H., Hepton, C.E.L. & Honey, S.I., 1979. 'The inshore benthic biota of the Lizard Peninsula south west Cornwall. I, The marine algae: history; *chlorophyta; phaeophyta*', *Cornish Studies*, 7, 7-37.

50    Price, J.H., Hepton, C.E.L. & Honey, S.I., 1979. 'The inshore benthic biota of the Lizard Peninsula south west Cornwall. I, The marine algae: *rhodophyta*; discussion', *Cornish Studies*, 8, 7-37.

51    Ralfs, J., 1880-5. 'Flora of Cornwall', MS in Penzance Subscription Library.

52    Rees, W.J. & Lumby, J.R., 1954. 'The abundance of octopus in the English Channel', *JMBA*, 33, 515-36.

53    Rostron, D., 1987. *Surveys of Harbours, Rias and Estuaries in Southern Britain. The Helford River. Report to the Nature Conservancy Council from the Oil Pollution Research Unit* (Field Studies Council, 69 pp.).

54    Short, F.T., Iblings. B.W. & Den Hartog, C., 1988. 'Comparison of a current eelgrass disease, to the wasting in the 1930s', *Aquatic Botany*, 30, 295-304.

55    Southward, A.J., 1980. 'The Western English Channel - an inconstant ecosystem?', *Nature*, 285, 361-6.

56    Spooner, G.M. & Holme, N.A., 1986. 'Studies on the marine life of the Helford River 2. Results of a survey in September 1949', *Cornish Biological Records*, 10, 1-29.

57    Smith, J.E., 1968. *Torrey Canyon Pollution and Marine Life. Report of the Marine Biological Association* (MBA/CUP).

58    Smith, S.M., 1991. '*Calyptraea chinensis*' (L. 1758) in Loch Ryan, *Porcupine Newsletter*, 5, 48-9.

59    Smith, S.M. & Heppell, D., 1991. *Checklist of British Marine Mollusca* (National Museums of Scotland Information Series, No. 11).

60    Swaby, S.E. & Potts, G.W., 1991. '*Record year for rare British fishes*', *Zoo.Cornwall & I.of Scilly*, 1, 12-13.

61    Tompsett, P.E., 1991. *Helford River Survey. Monitoring Report for 1990, No.3. A Report to the HVMCA Advisory Group* (CBRU).

62    Tompsett, P.E. & Turk, S,M, in prep. 'Invertebrates and fishes from the Helford River: check list in time phases'.

63    Townsend, M., 1967. 'The common limpet (*Patella vulgata*) as a source of protein)', *Folia Biologica*, 15, 343-51.

64    Turk, F.A., 1984 (for 1983). 'A study of the vertebrate remains from May's Hill, St Martin's', *Cornish Studies*, 11, 69-78.

65    Turk, S.M., 1971. *Seashore Life in Cornwall and the Isles of Scilly* (Truro: Bradford Barton).

66    Turk, S.M., 1976. 'Ecology notes on Helford Passage, Cornwall', *Conch.Newsletter*, 58, 512-15.

67    Turk, S.M., 1977. 'Compilation of notes and lists from the flats at Treath, Helford River'. Unpublished typescript.

68    Turk, S.M., 1979. 'Edward Step and the Long Drang: delving into the past and assessing the present with an eye to the future', *Conch.Newsletter*, 70, 159-62.

69    Turk, S.M., 1982. 'Influx of warm-water oceanic drift animals into Bristol and English Channels, Summer 1981', *JMBA*, 62, 487-9.

70    Turk, S.M., 1983. 'Cornish marine Conchology. Presidential Address to the Conchological Society', *J.Conch.Lond.*, 31, 137-51.

71    Turk, S.M., 1984. 'Helford Passage, April 1984, compared with 1972'. Unpublished typescript.

72  Turk, S.M., 1986. 'The three species of eelgrass (*Zostera*) on the
    Cornish coast', *Cornish Studies*, 14, 15-22.
73  Turk, S.M., 1988. 'Christopher Columbus and *Pteria colymbus*',
    *Conch. Newsletter*, 105, 93-5.
74  Turk, S.M., 1989. 'A new decline of *Zostera* in Cornwall',
    *Bot.Cornwall*, 3, 10-12.
75  Turk, S.M., 1990a. 'Further note on *Zostera* in Cornwall', *Bot.
    Cornwall*, 4, 18-19.
76  Turk, S.M., 1990b. *Helford River Survey. Monitoring Report for
    1989, No. 2. Report to the HVMCA Advisory Group by the CBRU.*
77  Wilson, D.P., 1947. 'The portuguese man-of-war, *Physalia physalis*
    L., in British and adjacent seas', *JMBA*, 27, 139-72.
78  Wilson, D.P., 1949. 'The decline of *Zostera* at Salcombe and its
    effect on the shore', *JMBA*, 28, 395-412.
79  Wilson, DP. & M.A., 1956. 'A contribution to the biology of
    *Ianthina janthina* L.', *JMBA*, 35, 291-305.
80  Wilson, D.P., 1958. 'On some small *Lanthina janthina* L. stranded
    on the Isles of Scilly, 1957', *JMBA*, 37, 5-8.

# THE EFFECT OF CHANGING CLIMATE ON MARINE LIFE: PAST EVENTS AND FUTURE PREDICTIONS

## A.J. Southward & G.T. Boalch

## Introduction

To detect changes in the natural environment and to predict future developments we need to know how much variation has occurred in the past. In the equivalent electronic terms we have to determine the amount of background noise that might conceal any emerging signal of change. We are fortunate in having records of biological changes during the last century, and much longer records of climate. For earlier periods, however, beyond the seventeenth century for climate and prior to the 1880s for marine life, only a limited amount of information is available. There are two main sources for the earlier data. One is historical documents that deal with climate, or climate effects or contain data on fish catches or fish cargoes. The other is chemical and biological measurements performed on the layers in core samples from ocean and lake sediments and the polar ice caps. The combined data from all sources tells us that the earth's climate has varied considerably in the last few million years and is still varying on a lesser scale today. It is against this record of fluctuating climate that we have to assess the recent biological data and derive estimates of the extent of any future change in response to changing climate. We first summarize the moderating effect of the sea on climate, then describe past changes and assess the influence of man's activities on climate and on marine life, concluding with some predictions of future events.

## Background: the sea as a moderator of climate

The sea occupies nearly three-quarters of the surface of the globe and is an important controlling factor in the temperature budget of the earth due to its great heat capacity.[1] It receives and stores a large part of the radiation from the sun. In both its physics and chemistry the sea is regarded as 'conservative' and many of the changes are slower and steadier than in the atmosphere.[2] As an example, in our latitudes, the seasons in the sea lag behind those on land, with minimum temperatures in February/March and the maximum in August.

Unlike the land there is little difference between day and night temperatures, and then only in the upper few cms in calm weather. In most parts of the ocean mixing caused by winds - i.e. wave action - carries the sun's heat downwards to a depth of 20 to 30 metres in coastal waters and more than a 100 metres in the open ocean.[3] At these limits there is a sharp boundary, the thermocline, and below this, in summer, the water is colder than at the surface. The difference can be as much as 5 to 10 degrees Celsius. There is very slow mixing across the thermocline by diffusion in both directions. In certain parts of the shallow seas where the tides are strong, and along the edges of the continental shelves and in places where the winds blow strongly offshore, there is much more mixing, involving upwelling of colder water. Figure 1 shows the detailed thermal structure in summer along a line from Roscoff to Plymouth, obtained with the undulating oceanographic recorder. The relative coolness of the surface water in the strongly tidal area off the north Brittany coast contrasts with the occurrence of a warmer surface layer south of the Eddystone off Plymouth where the water is usually stratified from May to August. In these latitudes in the autumn the sea loses heat to the air, the surface layer cools and the deeper layer warms up by diffusion across the thermocline, which then breaks down; the water remains mixed until the following spring when solar heating increases again. In contrast, in the tropics the surface waters are always warmer than the cold deep water. In the Arctic in winter the cooling is so intense that the surface layer becomes denser than the underlying water, resulting in its sinking towards the sea bed. Much of the cold deep water of the world oceans is formed in this way and circulates round the globe,[4] while there is a surface flow of warmer water circulating in the opposite direction (Figure 2) The sea thus acts as a global feedback system, performing functions akin to air conditioning and central heating. However, it is important to note that the present pattern of global circulation is not immutable; it is thought that different systems occurred in past epochs, and comparable changes might occur in response to changes in climate.[5]

## Evidence of past changes

There are better records of past changes in climate and in sea temperature than for the organisms that live in the sea, so the data is reviewed separately.

### *Changes in temperature*
The meteorologists are fortunate in having a long database of air temperature measurements, but air temperatures vary considerably

and there is still much discussion about trends. A well-known series on air temperature of Central England, illustrated in Figure 3 was originated by Professor G. Manley.[6] He extended the Meteorological Office data back from the 1800s to 1660 by drawing on private temperature measurements made by learned men of the time, as reported in their diaries and other records. Continuation of this series can be made from current meteorological station records.[7] There is much fluctuation in this record, but little doubt that conditions over most of England have been warmer in the present century than in the previous two centuries. By averaging records from a large number of meteorological stations some of the variations can be smoothed out.[8] Figure 4 shows the average temperature for the whole northern hemisphere since 1880, with an upward trend evident.

We are less fortunate for past details of sea temperature. Serious observation did not begin until the 1860s when merchant ships were asked to record weather observations for the Meteorological Office. There had been earlier attempts to observe ocean temperatures, of course, including the tracking of the Gulf Stream, for which there are some excellent US observations,[9] but sustained series were not then thought of. Some coastal stations started to keep weather records in the middle of the last century. Many seaside resorts had joined in by the 1890s, when it became popular to publicise tables of sunshine and weather, including sea temperature, reported each day. A sort of resort league was set up, still maintained today in the comparative reports of daily sunshine and other matters appended to the TV weather reports, with resorts vying with one another to appear the most favoured. The Plymouth series for inshore sea temperature was begun in the late 1890s, and other south-west resorts kept similar records. The different records do not always agree in the short term but because of the length of the series available they show a highly significant statistical agreement in their long-term trends and in their relationship to air temperature, indicating a common thread responsible for up to half of the observed variation.[10] In other words the pattern of weather, including the amount of solar radiation reaching us, gives similar temperature trends over most of the country and the nearby seas.

Figure 5 compares recent trends in average yearly sea temperature in Plymouth Sound, off Plymouth and in the nearby part of the Bay of Biscay. Small year to year variations (some of the noise) are filtered off by taking the running five year averages, and the longer-term trend is also indicated. For the first half of the century all the records show rising trends i.e. the sea was getting warmer, and so was the air (see Figure 3). This is particularly noticeable for the period from 1920 onwards, for which we are fortunate in having quantitative observations on marine life. The warming up was not steady but proceeeded in bursts which had a frequency comparable to the well-known sunspot cycle, lagging behind it by a year or so, as will be noted in more detail later. Compared with the period before 1920 or after 1960 the temperature conditions and the weather generally showed high stability - there were no great extremes of heat or cold. The weather pattern had a greater frequency of what the metereologists call 'westerly weather' which can be defined as a pressure pattern with a high to the south of the UK and a low to the north.[11] This weather pattern results more often in winds from the south west and south rather than the west, bringing warmth and contrasts with more northerly types of weather that bring cold winds from the west and north, such as occurred with greater frequency after 1961.

The cooler trend that set in after 1961, and continued until quite recently, is also indicated by the sea temperature records, but to differing degrees, conditions being cooler in Plymouth Sound, for example, than in the Bay of Biscay (Figure 5). The warm spell of 1920-60 shown by the temperature records was accompanied by quite marked changes in marine life in the English Channel and less obviously elsewhere around the U.K. Many of the changes were then reversed in the cooling phase that followed.

## Changes in marine life

In considering the influence of climate change on marine life it is important to note that most marine life, apart from whales and seals, whether they are fish or 'lesser' forms of life such as seaweeds, shellfish and crabs, or worms, are 'cold-blooded' i.e. their body temperature is the same as that of the sea around them. It is not possible for cold-blooded organisms to accommodate the global range of temperature, such as the 3 °C or more, experienced between the Arctic Ocean and the tropics: the biochemistry of the species is just not up to it. So there are different species occupying different climate zones, but otherwise playing a similar ecological role - there are, for example, arctic prawns, temperate prawns, warm-water prawns and tropical prawns for example. An analogy may be drawn with the division of garden plants into alpine species, hardy species, half-hardy species and stove/greenhouse species. In general then, it can be expected that the species that live in warm waters will do better in warmer years, and cold-adapted species should do better when the local climate is cooler.

It so happens that the western end of the Channel is close to a boundary between species of fish and invertebrates that flourish best in colder waters and those that do better in warmer waters. A good example of this is the cold-water herring on the one hand and the warm-water pilchard on the other (Figure 6). The ranges of these two clupeid fish overlap on our western coasts, but the extent of the over-lap is small compared with the area of sea occupied by each. Data on catches by sailing drifters at the turn of the century, before the days of mechanised fishing, illustrate the differing distributions of the two fish well.[12] Going up the Channel from the tip of W. Cornwall to E. Devon a predominance of pilchards gave way to predominance of her-rings. A comparable pattern of distribution is found in the common barnacles on the seashore. There is a cold-water species and a pair of warm-water species, the two sets overlappng along the south and west coasts.[13] These are two rather extreme examples of the way in which marine life relates to climate, and we have to visualize similar relation-ships as occurring throughout the ecosystem.

Our knowledge of changes in marine life in the South West comes from a number of sources. The longest series is for catches of marketable fish. There are official records of fish landings at the U.K. ports back to 1890 but the series is only reliable from 1903.[14] Prior to the official fishery statistics some records exist of export of certain fish products (e.g cured pilchards), and there are also nineteenth-century newspaper accounts of landings and catches. Farther back there are the Port Books, records kept for the Treasury, extending from Elizabethan times to the late eighteenth century, giving details of fish landings and exports, mostly those species subject to tax or subsidy. Finally, there are anecdotal records provided by published accounts of travellers and natural historians. For other marine life the evidence is less continuous but the intermittent records can be compared with the changes in the more consistent series for fish and plankton. As already noted, we do have detailed records of the occurrence and abundance of marine life in the South West during the present century. These records are among the serial observations maintained by the Marine Biological Association at Plymouth up to 1988 (Table 1). The longest and most quantitative records are those dealing with zooplankton and planktonic stages of fish, from 1924 to 1988. In some of the earlier years of the Association regular monitoring was carried out over the whole Channel and out into the nearest part of the North-ern Bay of Biscay, and in the 1960s and '70s the western part of the same area was sampled quite frequently. However, most of the records come from a group of stations close to Plymouth (Figure 7) the inner ones sampled weekly and the outermost at monthly intervals. Figure 8 gives some graphs from these data to illustrate the sort of changes that were found. What emerges most strongly from analysis of these records is that the biological changes were much more pronounced than the accompanying physical and chemical changes. The invertebrate animals of the plankton and the planktonic stages of fish showed increases or decreases of several orders of magnitude; in contrast sea temperature changed less than half a degree while changes in water chemistry did not exceed 30 per cent of the long-term average.

TABLE 1

The Plymouth Series of Environmental Observations

|  | from |
|---|---|
| Sea temperature and salinity | 1902 |
| Inorganic nutrients | 1921 |
| Organic nutrients | 1964 |
| Phytoplankton taken in nets | 1903 |
| Zooplankton | 1903 |
| Planktonic stages of fish | 1924 |
| Demersal fishes | 1913, 1920-2, 1950-1, 1976-9 |
| Intertidal barnacles | 1950 |

Our records show that after temperatures had begun to rise, and especially after 1930, many of the marine organisms with cold-water affinities became less abundant and the warm-water species became more abundant i.e. the boundary between the climate zones had been shifted towards the north. This change was particularly evident in the herring and pilchard fisheries. Off Plymouth the formerly flourishing herring fishery collapsed.[15] The herring shoals failed to maintain themselves, a situation not helped by the increasing fishing power of the fleets. Coincidentally, pilchards spread and became more abundant, breeding in large numbers in the waters off Devon and Cornwall.[16] By 1936 the Plymouth herring fishery was abandoned. Some herring fishing continued farther east in Lyme Bay, but finally petered out in 1950. We have been able to trace previous fluctuations in the relative abundance of these two fishes back to Elizabethan times[17] and show here some of the periods of dominance of each alongside the long-term trends in Central England air temperature (Figure 9). The air temperature series only goes back to the mid-seventeenth century. Historical records suggest there was a period of abundant pilchard about 1570-90. We can assume from European data for the timing of the wine harvests,[18] the tree-ring isotope data and sunspot cycle[19] that this too was a warm period, and an attempted reconstruction of the probable temperature is shown in Figure 9. In contrast, the severe cold spell of the second half of the seventeenth century, often called the 'little ice age', correlates with a severe reduction of the Cornish pilchard fishery and increases in abundance of herring in west Cornwall. Most recently, the failure of the herring fishery off Plymouth after 1930, and the coincident increase in abundance of pilchards[20] corresponded to a long period of higher temperatures.

Bottom-living fishes show changes in the same way as the pilchards and herrings that live in mid-water. Figure 10 compares the distribution of a warm-water fish, red mullet, with a cold-water fish, haddock; the range of other cold-water fish such as cod and ling would be similar. The MBA records show that many changes occurred in the species abundance of bottom fish such as these. The same grounds off Plymouth were sampled at different periods by 130 hauls of a similar size of trawl towed from a research vessel. The changes from the 1920s to the 1950s and from the 1950s to the 1970s are shown in Figures 11 and 12 as the log of the percentage difference, so as to compress the data into the graph. There is a group of seven species which are commoner in the warmer seas to the south of our area, and these fish increased in relative abundance between the 1920s and the 1950s during a period of warming climate; they declined again when temperatures fell after 1961. Another group of eight species, more abundant in seas to the northward of Devon, became less abundant after the climate warmed up, but returned again in higher proportion in the local catches after the climate began to cool. This latter group of fish comprise many of the marketable species most popular in England, and their scarcity in the 1940s and 1950s was a setback to the fishing industry in Devon and Cornwall. The revival of fishing at ports in the south-west since 1960 is sometimes said to have been due to greater investment in boats and to success in excluding foreigners from the inshore grounds (6 miles, then 12 miles off the coast), but the effect of climate change on species composition was the decisive factor.

It is not possible to be certain if there were natural fluctuations in bottom-living fishes previous to the present century. But Elizabethan records suggest that hake, a warm-water species, was co-abundant with pilchard during the warm spell before 1600.[21] So it is highly probable that species of bottom fish have also fluctuated in a manner comparable to the herrings and pilchards.

Similar changes occurred throughout marine life, but owing to the extent of interactions between the species there were changes in character of the ecosystem. During the period of warming we were headed towards a blue-water type of community, with clearer water, smaller plankton organisms and less food for some of the animals that had survived the change in temperature. This impression was strengthened by reports of tropical fishes being found, and by the increased catches of blue-sharks by sports anglers.[22] This period of increased temperatures and increases in warm-water animals was not immediately apparent at the time. Much of the change became obvious only in the 1950s, with a long series of records available. As noted, the change in mean temperature was only of the order of half a degree Celsius, say one degree Fahrenheit. However, on land this change would be equivalent to a difference of 300 feet in altitude, and a reverse change of this order would be enough to make much of the marginal hill farms untenable. On favourable slopes in the south of England it might be enough to make the difference between a viable and non-viable vineyard.

### Changes after 1961
When the changes in temperature and marine life became apparent in the late 1950s it was possible to make a prediction, which was that if things cooled down again the biota would change back to what had been found before 1930, that the cold-water species would increase and the warm-water species retreat.[23] Such a forecast was rendered more likely by consideration of some of the longer-term trends in climate. The good fit between the 11-year sunspot cycle and the 5-year smoothed averages of sea temperature between 1920 and 1960[24] indicated a connection with solar radiation. At the time such a connection was difficult to prove since there was no evidence of any linking mechanism between the sunspots, the weather, global temperature and biological changes. Recent studies show that the sunspots, areas of turbulence in the sun's exterior, can now be regarded as external evidence of real changes in the sun, matched by changes in solar radiation, neutrino flux and solar diameter.[25] In other words, more heat may reach the earth from the sun in the periods of higher sunspot activity. The 11-year cycle is just one of the evident solar cycles; higher harmonics have been detected, at 22, 45, 90 and 180 years. Comparable and longer period cycles have been detected in carbon isotope composition of tree rings. The amplitude of some of these longer period cycles appears to be related to movements of the other planets in relation to the sun, something that can be predicted. From this it appeared that the warming up that reached its peak in the late 1950s was part of such a longer cycle and that we could expect the following 45 years to usher in a period of cooling, presaged by lower peaks in the 11-year sunspot cycle. It has been established that even very small changes in solar radiation reaching the earth could have large effects on the climate.[26] A reduction of only 1 per cent would drop average temperatures by 5 °C.

The succession of cold winters and cool summers after 1961 indicated the end of the warming trend, and soon after there were changes in marine life. By 1972 it was evident that the trend for increases in cold-water forms was continuing, with reversals of the previous biological changes.[27] The fish data showed this remarkably well (Figure 12), with the warm-water species becoming less abundant and the cold-water species returning. Herrings did not return in much number however, when pilchards became scarce; their place was taken by large shoals of mackerel. Since 1980 the cooling trend has ceased and there is much evidence for increases in warm-water life again.

## Anthropogenic factors and climate change

Such was the cooling trend seen in higher latitudes in the 1970s that some pundits began to predict the start of another ice age. However, at the same time other experts began to detect signs that an opposite trend was possible and that global temperatures would rise as a result of human activities. Increased computing power allowed calculations of global mean temperatures, calculations could be corrected for errors including those due to unnatural warming of urban areas by human activities. Improved instrumentation, including gas chromatography, allowed more accurate measurements of atmospheric constituents of interest to climatologists and biologists. One of these measurements was of atmospheric carbon dioxide. In the late 1950s the now famous series of measurements of $CO_2$ began at Mauna Loa Observatory in Hawaii,[28] and a similar series was instigated at the US South Pole station. The value of series like these only becomes apparent after a number of years observations have accumulated, when trends can be seen above the noise of daily and monthly variations. The Hawaii measurements of atmospheric carbon dioxide concentration are shown in Figure 13. In the initial period all time series observations like these are at risk of being stopped, since the time span of research funding, especially from governments, is much shorter than the period needed to show a trend. It is fairly well-known that the Mauna Loa series was nearly stopped after a few years. The famous observations of the ozone hole over the Antarctic - another story, for which there is no space here - and currently the pride of the U.K. government, were very nearly stopped at an early stage by agents of that very same government. The observations of atmospheric $CO_2$ were expressly designed to determine whether there was a sustained increase due to burning of fossil fuels and other human activities. Carbon dioxide gas is not transparent to all infra-red radiation, and an increase in its concentration in the atmosphere would lead to greater trapping of the longer wavelengths of solar radiation that would otherwise be reradiated back into space - this is the so-called 'greenhouse effect' and an increase in the level of $CO_2$ in the atmosphere would be expected to increase the temperature of the earth.[29]

It was evident by the 1970s that the atmospheric concentration of $CO_2$ was increasing (Figure 13) and new instrumentation for gas chromatography showed that atmospheric concentrations of other infra-red absorbing compounds were also increasing, notably methane, the chlorofluorocarbons, low-level ozone and nitrous oxide.[30] These other compounds are present in very much lower quantities than carbon dioxide but their infra-red absorbing powers are greater and it is calculated that the amounts added to the atmosphere in the 1980s had a combined effect equivalent to the total $CO_2$ added in the same decade. Most of the rise in the 'greenhouse gases' can be attributed to human activities, and the amount added to the atmosphere from this source is still rising exponentially (Figure 14). Hence the deduction that global temperatures will rise is now seen to be scientifically correct, and the only question is when will we see the effects.

The temperature of the earth has varied on a geological time scale, witness the past traces of glaciers and fossils of arctic plants in Britain. Further back still, fossil evidence shows tropical conditions existing. Such changes can now be quantified from cores taken in deep-sea sediments and the polar ice-caps. Ageing can be carried out by isotope analysis e.g. of carbon and lead. Measurement of deuterium isotope ratios or oxygen isotope ratios, which vary with temperature, gives an insight into past climates; and most recently, analysis of air bubbles frozen in arctic and antarctic ice,[31] reveals the atmospheric $CO_2$ content of the time (Figure 15). Thus the 1986 'Vostok' ice-core[32] shows shows how temperature and $CO_2$ content have been synchronised quite well over the past 35,000 years (Figure 16). Other reconstructions can take the time scale back to the cretaceous epoch, when tropical conditions existed over Britain (Figure 17). In these remote periods, the data provided by cores suggests the ocean circulation pattern may have differed from that of today.[33] The evidence showing the synchrony between past natural changes in climate and atmospheric carbon dioxide, reinforces the view that anthropogenic rises in carbon dioxide will lead to rising temperatures and changes in marine and terrestrial life.

**Predicting climate change and biological effects**

It is possible to make some predictions about what will happen to climate and marine life in the next century based on the past evidence of change. Let us assess the propects for climate change and global warming first, then look at the probable effect on marine life.

*Possible changes in climate*

The cooling trend that set in after 1960 was obvious only in the northern hemisphere at higher latitudes. Elsewhere it was less marked and when global averages are examined the fall was slight and overshadowed by a strong upward surge in average temperature since 1970 (Figure 18). The recent warming has become apparent also in the sea; a return of warm-water species was detected in our data for the English Channel, apparently related to ocean temperature rather than coastal temperature (Figure 19). In the past few years we have seen unprecedented warming and increased occurrence of warm-water life in British waters. This change goes against the prediction made in the 1960s of a longer cooling trend. The question arises are we already seeing, superimposed on normal trends, an effect that could be attributed to global warming resulting from accumulation of greenhouse gases? The answer seems to be a qualified perhaps. But there is no doubt that current land and sea temperature measurements, when compared with the long time-series available, are showing changes quite outside past variation, and these changes appear to be in synchrony with carbon dioxide changes.[34]

In predicting the future extent of warming we have to consider the extent of feedback mechanisms, especially those built into the global carbon cycle, and the role played by carbon dioxide in this cycle. Carbon dioxide constitutes about 70 to 80 per cent of the accumulated greenhouse gases in the atmosphere. What happens to this increase in carbon dioxide is the subject of much current research and modelling.[35] First, a large part of the global carbon is present in the sea (Figure 20). Below the surface waters the amount of carbon dioxide increases with depth (and hence pressure), and the ocean depths thus constitute a tremendous reserve. Many scientists believe that if the deep waters of the oceans warmed up from the present very low temperature of 1 to 2 °C to surface temperature, and if they equilibrated with the atmosphere, the result would more than double the present atmospheric concentration of $CO_2$. This would in turn increase global temperatures, causing yet more release of carbon dioxide from the oceans, a positive feedback response which is thought by some alarmists to have the potential to 'runaway'. A positive feedback would also be involved if rising temperatures increased the amount of water vapour in the atmosphere, since water vapour also produces a greenhouse effect. Large ice-sheets like those on the poles and on high latitude mountain ranges constitute yet another positive feedback mechanism in that when they are extending they add to the earth's albedo, reflecting more solar radiation and thus reducing warming. Conversely, when they shrink they increase the amount of heat retained by the earth, accelerating their own shrinking.

There are also biological feedback mechanisms that can influence the global carbon cycle. One of these involves chemical changes. Many marine animals and plants lay down shells or skeletons composed of calcium carbonate, carbon dioxide combined with lime. These shells are formed at the ocean surface and in continental waters. On death of the organisms the shells fall to the bottom and accumulate in the sediments, thus removing some of the carbon from the surface waters, a negative feedback mechanism. Another negative feedback mechanism is related to the basic cycle of production in the sea. At the base of the supply of food to most life in the sea are the minute floating plants of the plankton, the phytoplankton. There are immense numbers of these, especially in spring and summer. Using energy from the sun, mediated by chlorophyll, like the green plants on land, they fix carbon dioxide into organic compounds. These plants are eaten by animals in the sea, and when both die and decay the carbon in their bodies falls to the seabed and accumulates in the sediments. The extent to which plant life can take up carbon dioxide is well-established and recent expeditions have shown that in the North Atlantic local dense blooms of phytoplankton can reduce the relative amount of dissolved carbon dioxide. It is difficult at present to determine how much of this phytoplankton carbon becomes incorporated in the sediments so the extent of the negative feedback mechanism is unknown. Another much discussed biological feedback mechanism is positive. As measured by existing methods involving fixation of $^{14}C$ labelled bicarbonate, the annual consumption of $CO_2$ by microscopic plants of the plankton is highest in temperate latitudes and least in the tropics The bigger phytoplankton, especially the large diatoms that grow in colder waters, remove a lot of $CO_2$ from the surface water and much of it falls to the sea bed as organic matter, which is incorporated in the sediments and thus remove carbon from the global cycle. If the sea warms up these larger phytoplankton will tend to be replaced by smaller species, as happens now in summer.[36] The small phytoplankton undergo rapid recycling as food of animals, and less of the carbon they produce is thought to fall to the sea bed to be incorporated in the sediments. If - and this is still only hypothetical - there is less overall fixation of $CO_2$ when the climate warms up, then less $CO_2$ will be removed from the atmosphere than before, enhancing the greenhouse effect.

It has been remarked that if positive feedbacks operated in the past then it is difficult to explain the occurrence of both the ice ages and the warm interglacials (see Figures 16 & 17). Hence we have to postulate that negative feedbacks occur which will have to be incorporated in the models. Such, for example, might be the increasing calcification of many organisms, including algae, in warmer water, giving greater rather than lesser removal of $CO_2$ from the global cycle. The enormous deposits of calcium carbonate along the Channel Coasts e.g. the chalk cliffs of Dover and the opposite side, are uplifted marine sediments, mostly laid down in shallow warm seas in which giant reptiles flourished. Much of the chalk comes from the minute bodies of small phytoplankton organisms - the coccolithophores. These plants flourished in warm seas where the $CO_2$ level has been estimated from isotope studies to be as high as 800 ppm,[37] a level which they must have helped to reduce by negative feedback.

Thus, although we can predict what is called the climate forcing effect of the carbon dioxide produced by human activities, we cannot easily calculate the degree of feedback. This introduces much uncertainty in the quantification of the changes in temperature that might be produced. Any predictions need to take into account this uncertainty. Figure 21 shows predictions of temperature change that might result from the greenhouse effect, with and without feedback. This graph should be viewed with reserve since not all feedback mechanisms can be taken into account, but it does indicate a potential rise of between 1 and 3 °C in global averages in the next fifty years. If temperatures rise to this extent then it is predicted that sea-level will rise, by combination of expansion of sea-water as it gets warmer and by melting of mountain glaciers.[38] Observations of global sea level show it is already rising, from a combination of causes not necessarily related to human activities (Figure 22). If the rise in sea level accelerates much coastal wetlands might be permanently flooded with detrimental effects to marine and coastal wildlife as well as to sea defences and cities. Figure 22 also shows an estimate of greenhouse effect-driven rise in sea level for one part of the world; the biological consequences of this change are discussed in the next section.

### Predicting biological response to warming

The biological records show that a small rise of sea temperature of the order of half a degree between 1920 and 1960 was connected with quite obvious changes in marine life. Some of these changes reversed during a period of cooling from 1962 to 1980. We can use these events to predict what might happen as a result of further changes in temperature. If, as a result of the greenhouse effect, sea temperatures rise, then the recent increase in southern warm-water species detected since 1980 will continue and may go beyond what was observed in the previous peak years of warming from 1940 to 1960. We could see development in the western English Channel of a plankton community similar to that found in the blue-water regions of the southern Bay of Biscay. Changes will be occur at all levels in the ecosystem, but if the rate of change is very rapid then some instability will develop. If the ecosystem becomes less stable and if small phytoplankton such as dinoflagellates and flagellates dominate for longer periods each year then there is a risk of increased occurrence of 'red tides' inshore, with deleterious consequences to the shallow-water life, including shellfish and fish, and also to tourism.

The most obvious change is expected to be a poleward migration of fish species,[39] bringing a greater abundance of warmer-water species to the Channel, for example hake, black bream, red mullet and, of course, pilchard. The typical cold-water species such as cod, haddock, ling, plaice and lemon sole would decline or be forced to retreat to the north. Such a replacement of cold-water fish by warm-water fish of lesser market appeal would need changes in the fishing industry and in the attitude of consumers, English and otherwise. We might also see the final decline of the English oyster, already hard-pressed by pollution. In contrast, if the sea warms up by a few degrees then the Portuguese or Lusitanian oyster and maybe the Japanese oyster may find they can breed in British waters. We might also find increases in the spiny lobster, to offset the present appalling decline of the ordinary lobster, which is severely overfished at present in the South West. An increased abundance of pilchards would benefit the fish meal industry, which could also make use of probable increases in horse-mackerel and other fishes not of much market appeal at present. On the debit side, loss of marine wetlands to rising sea level along the coast would be a setback to species that use these places as nursery grounds, for example soles and flounders. Birds and other wildlife that live in or visit these wetlands would also suffer.

## Conclusions

The effect of rising temperatures would result in many changes in marine life but the picture is not all gloomy. We believe that scare scenarios of global warming that include a 'runaway' effect, reveal a lack of real ecological knowledge on the part of the propounders. That negative feedbacks must exist in the global carbon cycle, evident from the past record of the rocks and sediments, was noted as long ago as 1948.[40] Ultimately the ecosystem has the ability to cope with increases in global carbon dioxide and achieve stability. But changes there will be, of the sort we have been suggesting. One possible view, not necessarily that of any official body, is that we are already seeing an effect of global warming on the marine ecosystem, but that the trend due to this has been superimposed on a more 'natural' oscillation of the system, possibly related to medium and long-term cycles in solar activity. Only very detailed data, such as that obtained by the Marine Biological Association during the present century, is capable of showing up trends when there are many overlapping factors of importance. We need to resume monitoring of the environment and marine life in the South-West, to follow the expected changes.

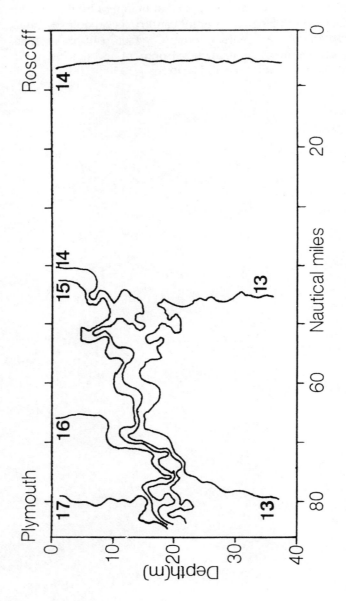

Figure 1. Summer sea temperatures in the western English Channel on a line from Roscoff in Brittany to Plymouth. A very detailed plot of temperature was recorded virtually continuously by the 'undulating oceanographic recorder' towed behind a Brittany Ferries vessel on a routine run. The recorder takes an oscillating path, moving from the surface to 40 metres deep and then back again repetitively. On the Brittany side the water is mixed by strong tides and is all about 14 °C; on the Plymouth side the water is stratified, warm at the surface and cold below (data from J. Aiken, Plymouth Marine Laboratory).

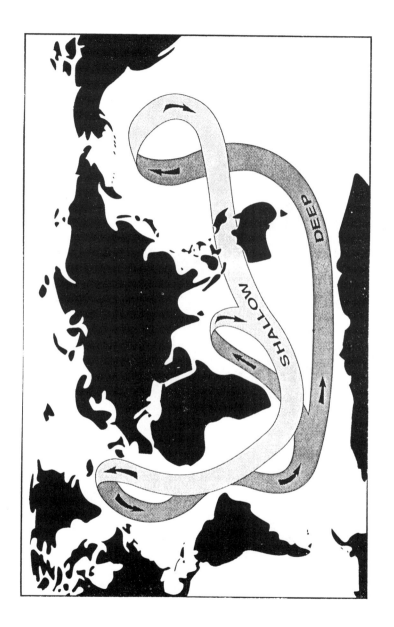

Figure 2. Generalised sketch of the main ocean circulation pattern today.[41]

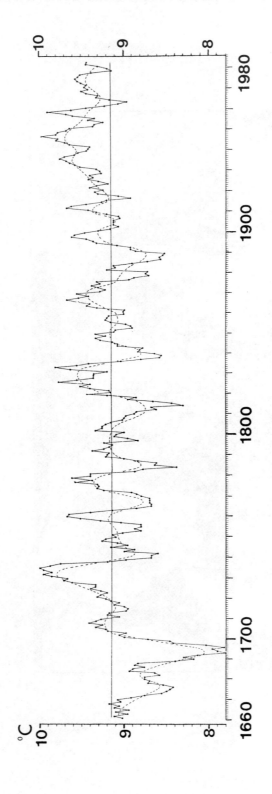

Figure 3. Air temperatures over Central England from 1660 to the present. Shown as 5-year smoothed averages (solid line) and longer-term trends (broken line).[42]

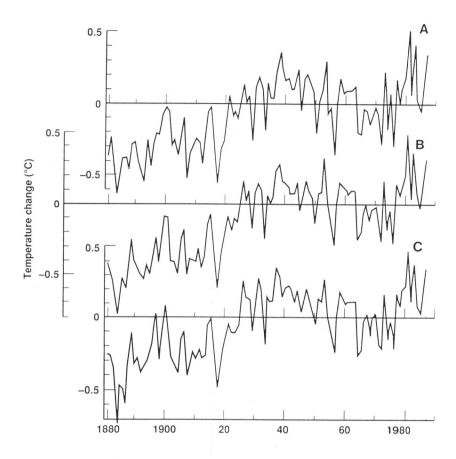

Figure 4. Corrected air temperatures for the whole Northern Hemisphere, from three different methods of correction, all using the same raw meteorological records.[43]

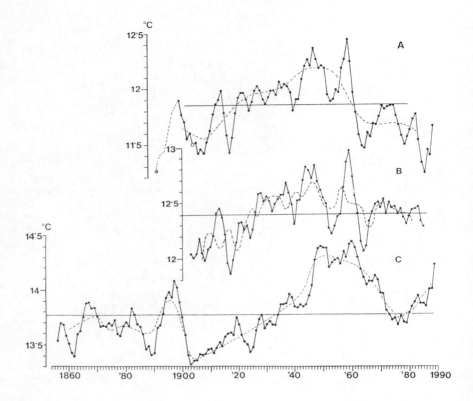

Figure 5. Sea surface temperatures: A) in the northern Bay of Biscay[44]; B) in the Channel off Plymouth[45]; C) in Plymouth Sound.[46] The solid lines show 5-year smoothed averages, the broken lines are longer-term trends.

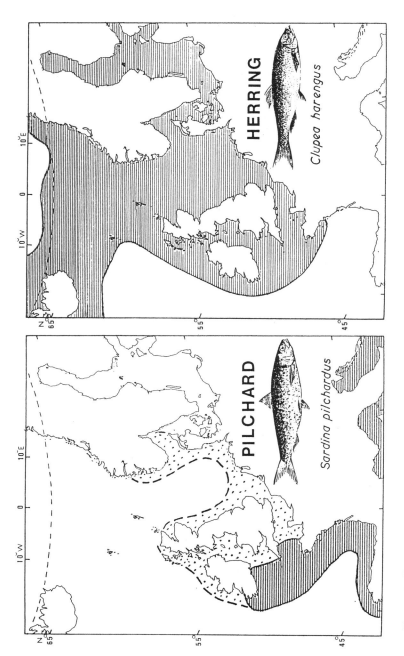

Figure 6. The distributions of herring and pilchard compared. The pilchard breeds over the area shown hatched but occurs over a wider range, shown dotted. The data for herring refers to conditions fifty years ago and the present exact southern limits are not fully known.

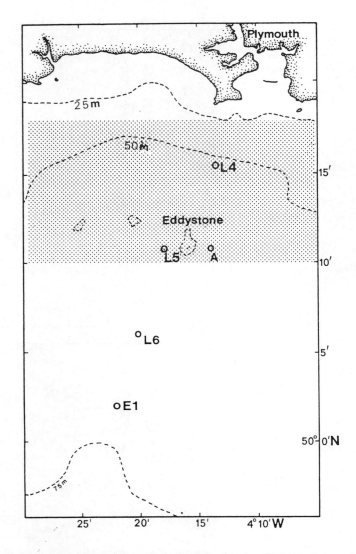

Figure 7. The routine stations sampled by the Marine Biological Association until 1987, when Research Council (government) support was withdrawn. E1 was sampled monthly, L5 or A weekly, L4 more often. The stippled area denotes the region from which bottom-living fish were sampled and from which net phytoplankton samples were taken several times a week.

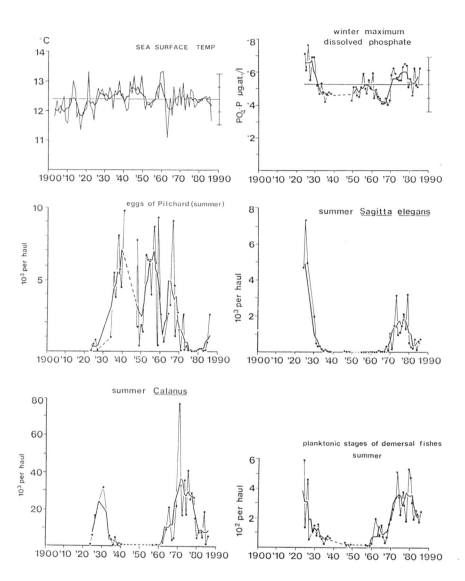

Figure 8.   Examples of some of the routine records from the Plymouth programme: the annual values are shown as dots and thin lines, the five-year smoothed values as heavy lines.

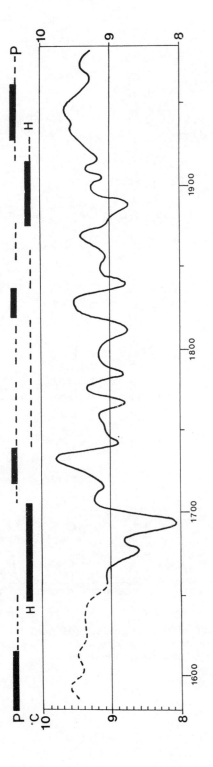

Figure 9. The long-term trends in Central England air temperatures, shown as deviations from the mean for the whole period and used as a guide to sea temperature off Plymouth. Periods when the records, including historic data, show the abundance of pilchard (p) and herring (h) are placed above the temperature trend. Temperatures from before 1650 have been visualised by comparison with sunspot and tree-ring data; from contemporary references to weather, especially the stormy 'Armada' summer of 1588, Lamb[47] believes the onset of the 'little ice age' began after 1550, but the fish and sunspot/tree-ring data indicate a later onset.

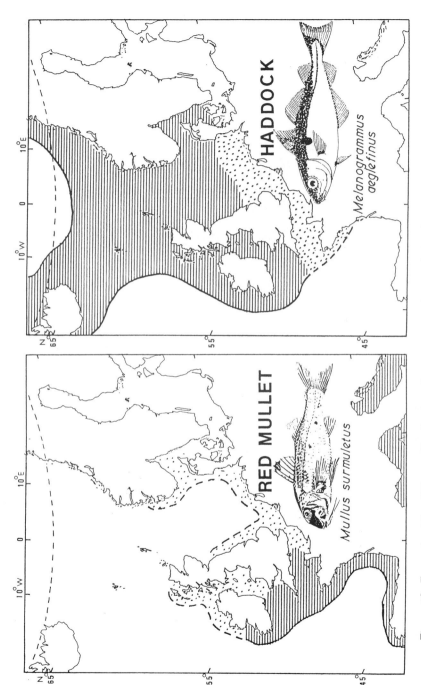

Figure 10. Example of the different distribution of a warm-water bottom-living fish (red mullet) compared with a cold-water fish (haddock). The region of continuous occurrence is shown hatched, of occasional occurrence dotted.

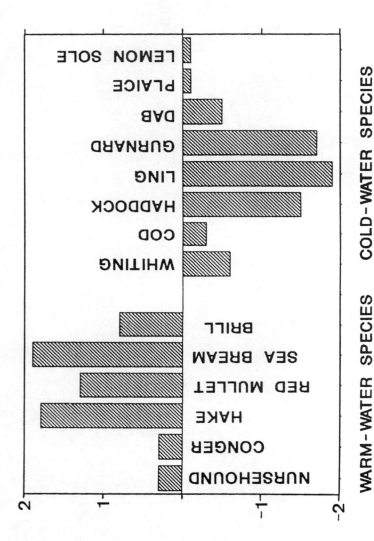

Figure 11. Changes in the proportions of certain bottom-living fishes in catches of research vessels between 1919-22 and 1950-52. The difference is expressed as the log of the percentage increase or decrease (Marine Biological Association data).

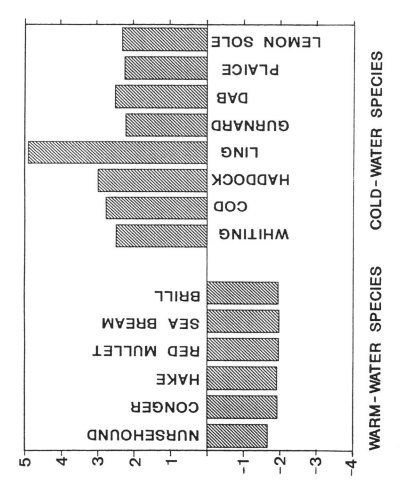

Figure 12. The same fish as shown in Figure 11, but comparing the change in catch from 1950-52 to 1976-79.

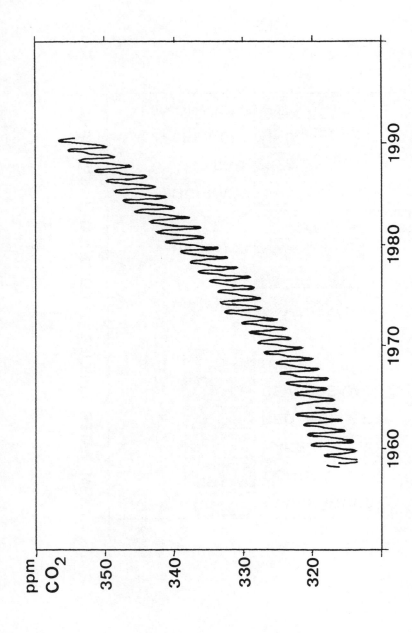

Figure 13. The concentration of carbon dioxide in the atmosphere at Hawaii as parts per million. The monthly mean values have been joined together into a continuous graph. There is an annual increase and decrease each year, related to seasonal growth of vegetation, but the overall rising trend is now clear.[48]

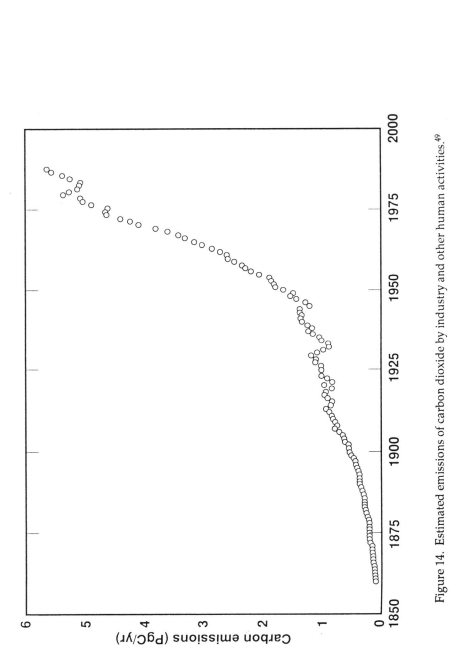

Figure 14. Estimated emissions of carbon dioxide by industry and other human activities.[49]

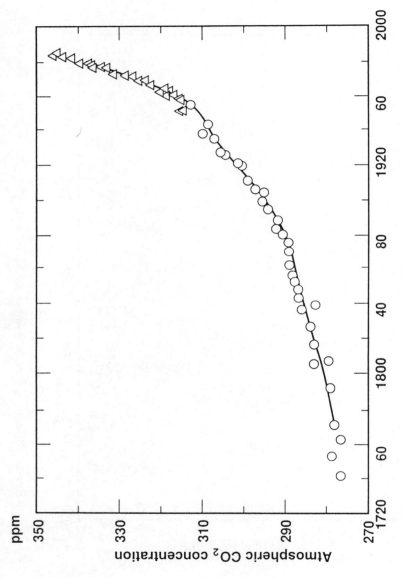

Figure 15. Extending values for the amount of carbon dioxide in the atmosphere back into last century by analysis of air bubbles frozen in the ice cap. The recent trend from 1960 is taken from the data used in Figure 13.[50]

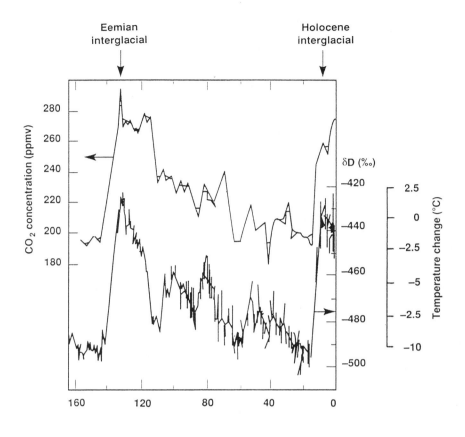

Figure 16. Comparison of the 160,000 year record of carbon dioxide concentration (air bubbles) and temperature (deuterium isotope ratio of the ice) obtained from a core ('Vostok') taken from the Antarctic ice cap.[51]

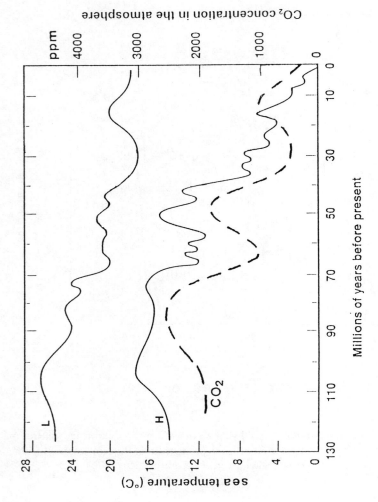

Figure 17. A long-term estimation of atmospheric carbon dioxide concentration and sea temperatures in high (H) and low (L) latitudes, from isotope ratios of shells in deep-sea sediment cores.[52]

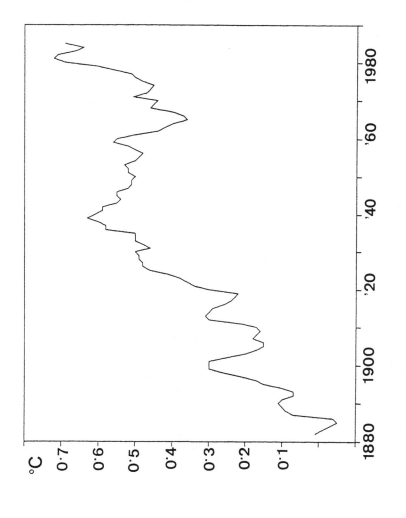

Figure 18. Estimated trend in global air temperature since 1880.[53]

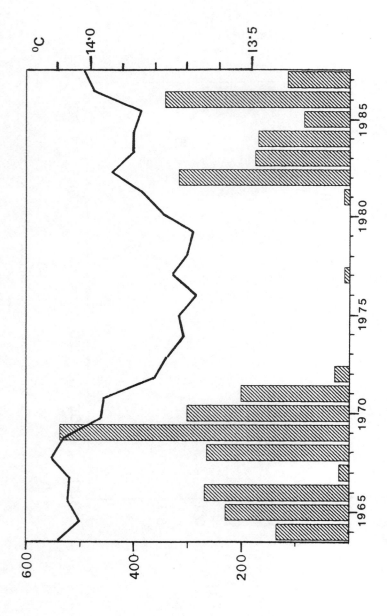

Figure 19. Abundance of a warm-water copepod (*Eucalanus*) in the autumn plankton off Plymouth (Marine Biological Association data) compared with smoothed averages of sea temperature in the nearby part of the Bay of Biscay.

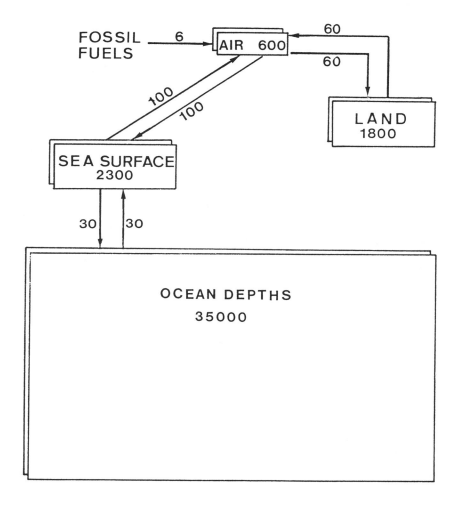

Figure 20. The relative amounts of the global carbon held in the atmosphere, the oceans and the land in gigatonnes, showing also the relative extent of the exchanges between these categories and the annual input from human activities.[54]

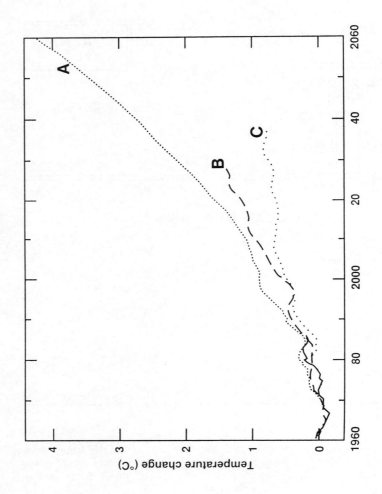

Figure 21. Recent trends in global air temperature extended into the next half century, using three different models: A) assumes emissions of greenhouse gases continue to grow at the present rate; B) assumes emissions cease to increase after the year 2000; C) assumes the rate of increase falls to zero by 2000.[55]

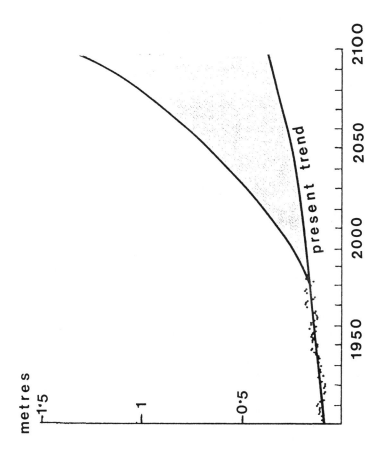

Figure 22. The present trend of rising sea level and an estimate of the projected rise next century as a result of global warming.[56]

## NOTES

1   M.I. Budyko,*Climate and Life* (transl. & ed. D.H. Miller) (Academic Press,1974); M.I. Budyko, *The Evolution of the Biosphere* (D. Reidel, Dordrecht,1986).

2   H.W. Harvey, *The Chemistry and Fertility of Sea Waters* (Cambridge, 1955).

3   H.U. Sverdrup, M.W. Johnson & R.H. Fleming, *The Oceans. Their Physics, Chemistry and General Biology* (Prentice-Hall, Englewood Cliffs, N.J., 1942).

4   W.S. Broecker, D.M. Peteet, & O. Rind, 'Does the ocean-atmosphere system have more than one stable mode of operation?', *Nature, Lond.*, 315, 21-6 (1985); W.S. Broecker, 'Unpleasant surprises in the greenhouse?', *Nature, Lond.*, 328, 123-6 (1987).

5   W.S. Broecker, D.M. Peteet, & O. Rind, 'Does the ocean-atmosphere system have more than one stable mode of operation?', *Nature, Lond.*, 315, 21-6 (1985).

6   G. Manley, 'Central England temperatures: monthly means from 1659 to 1973', *Quarterly Journal of the Royal Meteorological Society*, 100, 389-405 (1974).

7   H.H. Lamb, *Climate. Present, Past and Future. 2. Climatic History and the Future* (Methuen, 1977); H. H. Lamb, *Weather, Climate and Human Affairs* (Routledge, 1988).

8   P.D. Jones, R.S. Bradley, H.F. Diaz, P.M. Kelly & T.M.L. Wigley, 'Northern Hemisphere surface air temperature variations: 1851-1984', *Journal of Climate and Applied Meteorology*, 25, 161-79.

9   M.F. Maury, *The Physical Geography of the Sea*, originally published in 1855, reprinted with notes by J. Leighley, ed., (Belknap Press of Harvard University, 1963).

10  A.J. Southward, G.T. Boalch & L. Maddock, 'Fluctuations in the herring and pilchard fisheries of Devon and Cornwall linked to change in climate since the 16th century', *Journal of the Marine Biological Association of the United Kingdom (JMBA)* 68, (1988) 423-45.

11  L. Maddock & C.L. Swann, ' A statistical analysis of some trends in sea temperature and climate in the Plymouth area in the last 70 years', *JMBA*, 57 (1977) 317-38.

12  E.S. Russell, 'Report on log-book records relating to mackerel, pilchards and herring, kept by fishermen during the years 1896-1911 under the auspices of the Cornwall County Council', *Fishery Investigations. Ministry of Agriculture, Fisheries and Food* (ser.2), 3(1) (1915); see also Figure 2 in Southward, Boalch & Maddock,

'Fluctuations in the herring and pilchard fisheries', *JMBA*, 68 (1988) 423-45.

13  D.J. Crisp & A.J. Southward, 'The distribution of intertidal organisms along the coasts of the English Channel', *JMBA*, 37, 157-208 (1958); A.J. Southward, 'Forty years of changes in species composition and population density of barnacles on a rocky shore near Plymouth'. *JMBA*, 71, 495-513 (1991).

14  *Annual Reports of Proceedings under Acts relating to Sea Fisheries (England and Wales)*, Board of Agriculture (HMSO, 1903-18); *Sea Fisheries Statistical Tables (England and Wales)*, Ministry of Agriculture and Fisheries (HMSO, 1919-66); *Monthly Returns of Sea Fisheries Statistics (England and Wales)*, Ministry of Agriculture, Fisheries and Food (MAFF Fisheries Statistics Unit, 1967-86).

15  D. H. Cushing, 'On the failure of the Plymouth herring fishery', *JMBA*, 41, 799-816 (1961).

16  D.H. Cushing, 'The number of pilchards in the Channel', *Fishery Invest., Lond.*, Ser. 2, 21(5) (1957), 1-27; A.J. Southward, 'The distribution of some plankton animals in the English Channel and Approaches', Pt. 2, *JMBA*, 42, 275-375, Pt. 3, *JMBA*, 43, 1-29.

17  A.J. Southward, G.T. Boalch & L. Maddock, 'Climatic change and the herring and pilchard fisheries of Devon and Cornwall', in D. Starkey, ed., *Devon's Coastline and Coastal Waters* (Exeter U.P., 1988), 33-57.

18  E. Le Roy Ladurie, *Times of Feast, Times of Famine*, (Doubleday, New York, 1971); H.H. Lamb, *Weather, Climate and Human Affairs* (Routledge, London, 1988).

19  J.A. Eddy, 'The Maunder minimum', *Science, N.Y.*, 192, 1189-1202 (1976); M. Stuiver & P.D. Quay, 'Changes in atmospheric carbon-14 attributed to a variable sun', *Science, N.Y.*, 207, 11-19 (1980); M. Stuiver & T.F. Braziunas, 'Tree cellulose isotope ratios and climatic change', *Nature, Lond.*, 328, 50-60 (1987).

20  A.J. Southward, G.T. Boalch & L. Maddock, 'Climatic change and the herring and pilchard fisheries of Devon and Cornwall', in D. Starkey, ed., *Devon's Coastline and Coastal Waters* (Exeter U.P., 1988), 33-57.

21  R.N. Worth, *History of Plymouth* (Plymouth, Brendon & Sons, 1890); R.N. Worth, *Calendar of the Plymouth Municipal Records* (Plymouth, Brendon & Sons, 1893).

22  F.S. Russell, 'The English Channel', *Transactions of the Devonshire Association*, 85, 1-17.

23  A.J. Southward, 'The distribution of some plankton animals..Pt. 3', *JMBA*, 63, 1-29.

24    A.J. Southward, E.I. Butler & L. Pennycuick, L., 'Recent cyclic changes in climate and in abundance of marine life', *Nature, Lond.*, 253 (1975), 714-17.

25    D.V. Hoyt, 'Variations in sunspot structure and climate', *Climatic Change*, 2, 79-92 (1979); J.A. Eddy, R.L. Gilliland & D.V. Hoyt, 'Changes in the solar constant and climatic effects', *Nature, Lond.* 300, 689-93 (1982); D. Gough, 'What causes the solar cycle?', *Nature, Lond.*, 319, 263-4 (1986); D. Gough, 'Deep roots of solar cycles', *Nature, Lond.* 336, 618-19 (1988); N.E. Newell, R.E. Newell, J. Hsiung & Z-X Wu, 'Global marine temperature variation and the solar magnetic cycle', *Geophysical Research Letters*, 16, 311-14 (1989); R.C. Willson & H.C. Hudson, 'The sun's luminosity over a complete solar cycle', *Nature, Lond.*, 351, 42-4 (1991).

26    M.I. Budyko, *Climate and Life* (transl. & ed. D.H. Miller) (Academic Press, 1974); H.H. Lamb, *Climate. Present, Past and Future. 2. Climatic History and the Future* (Methuen, 1977); J.A. Eddy, R.L. Gilliland & D.V. Hoyt, 'Changes in the solar constant and climatic effects', *Nature, Lond.* 300, 689-93 (1982); M.I. Budyko, *The Evolution of the Biosphere* (D. Reidel, Dordrecht, 1986).

27    A.J. Southward, 'The Western English Channel - an inconstant ecosystem?' *Nature, Lond.*, 285, 361-6 (1980); A.J. Southward & G.T. Boalch, 'Aspects of long-term changes in the ecosystem of the western English Channel in relation to fish populations', in *Long- Term Changes in Fish Populations*, T. Wyatt & M.G. Larraneta, eds, (Vigo, 1988), 415-47.

28    C.D. Keeling, R.B. Bacastow, A.E. Bainbridge, C.A. Ekdahl, P.R. Guenther, L.S. Waterman & J.F. Chinn, 'Atmospheric carbon dioxide variations at Mauna Loa Observatory, Hawaii', *Tellus*, 28, 538-51 (1976); C.D. Keeling, *Atmospheric $CO_2$ Concentrations - Mauna Loa Observatory, Hawaii 1958-1986* (Environmental Sciences Division, U.S. Department of Energy, publ. 2798, 1986).

29    S.H. Schneider, 'Climate modelling', *Scientific American*, 256(5), 72-81 (May, 1977); J. Hansen, D. Johnson, A. Lacis, S. Lebedeff, P. Lee, D. Rind and G. Russell, 'Climatic impact of increasing atmospheric carbon dioxide', *Science, N.Y.*, 213, 957-66 (1981); B. Bolin, ed., *Carbon Cycle Modelling* (J. Wiley, New York, 1981); A. Crane & P. Liss, 'Carbon dioxide, climate and the sea', *New Scientist*, 108 (1483) (1985), 50-3; J.R. Trabalka & D.E. Reichle, eds, *The Changing Carbon Cycle. A Global Analysis* (Springer-Verlag, New York, 1986); A. Henderson-Sellers, 'Climate is a cloudy issue', *New Scientist*, 115 (1570) (1987), 37-9; C. Sear, 'Wetter weather linked to greenhouse effect', *New Scientist*, 114 (1568) (1987), 27; S.H. Schneider, 'The greenhouse effect: science and

policy', *Science, N.Y.*, 243, 771-81 (1989); M.C. McCracken, M.I
Budyko, A.D. Hecht, A.D. & Y.A. Izrael, eds, *Prospects for Future
Climate* (Lewis Publishers, Chelsea, Mich,1990); M.C. McCracken
(chairman), *Energy and Climate Change. Report of the DOE Multi-
Laboratory Climate Change Committee* (Lewis Publishers, Chelsea,
Mich., 1991).

30    D.A. Lashof & D.R. Ahuja, 'Relative contributions of greenhouse
gas emissions to global warming', *Nature, Lond.* 344, 529-32
(1990).

31    A. Neftel, E. Moor, H. Oeschger & B. Stauffer, 'Evidence from
polar ice cores for the increase in atmospheric $CO_2$ in the past two
centuries', *Nature, Lond.* 315, 45-7 (1985).

32    J.M. Barnola, D. Raynaud, Y.S. Korotkevich & C. Lorius, 'Vostok
ice core provides 160,000-year record of atmospheric $CO_2$',
*Nature, Lond.*, 329, 408-14; Chappellaz, J.M. Barnola, D. Raynaud,
Y.S. Korotkevich & C. Lorius, 'Ice core record of atmospheric
methane over the past 160,000 years', *Nature, Lond.*, 345, 127-31;
J.R. Petit, L. Mounier, J. Jouzel, Y.S. Korotkevich, V.I. Kotlyakov
& C. Lorius, 'Palaeoclimatological and chronological implications
of the Vostok core dust record', *Nature, Lond.*, 343, 56-8 (1990).

33    W.S. Broecker, D.M. Peteet, & O. Rind, 'Does the ocean-
atmosphere system have more than one stable mode of operation?', *Nature, Lond.*, 315, 21-6 (1985); R.G. Fairbanks, 'A 17,000-
year glacio-eustatic sea level record: influence of glacial melting
rates on the Younger Dryas event and deep-ocean circulation',
*Nature, Lond.*, 342, 637-42 (1989); J.P. Kennett & L.D. Stott, 'Abrupt
deep-sea warming, palaeoceanographic changes and benthic
extinctions at the end of the Palaeocene', *Nature, Lond.*, 353, 225-9
(1991).

34    D. Rind, A. Rosenzweig & C. Rosenzweig, 'Modelling the future:
a joint venture', *Nature, Lond.* 334, 483-6 (1988); C. Kuo,
C. Lindberg & D.J. Thompson, 'Coherence established between
atmospheric carbon dioxide and global temperature', *Nature,
Lond.*, 343, 709-13 (1990); M.E. Schlesinger & X-J. Jiang, ' Revised
projection of future greenhouse warming' *Nature, Lond.* 350, 219-
23 (1991); C.S.M. Doake & D.G. Vaughan, 'Rapid disintegration of
the Wordie ice shelf in response to atmospheric warming', *Nature,
Lond.*, 350, 328-30 (1991).

35    T. Takahashi, 'The carbon dioxide puzzle', *Oceanus*, 32 (2), 23-9
(1989).

36    L. Maddock, G.T. Boalch, & D.S. Harbour, 'Populations of
phytoplankton in the western English Channel between 1964 and
1974', *JMBA*, 61 (1981), 565-83; G.T. Boalch, 'Changes in the

phytoplankton of the Western English Channel in recent years', *British Phycological Journal*, 22, 225-35; L. Maddock, D.S. Harbour & G.T. Boalch, 'Seasonal and year-to-year changes in the phytoplankton from the Plymouth area, 1963-1986', *JMBA*, 69 (1989), 229-44.

37    G.H. Rau, T.Takahashi & D.J. Des Marais, 'Latitudinal variations in plankton; implications for $CO_2$ and productivity in past oceans', *Nature, Lond.*, 341, 516-18 (1989).

38    J.L. Jacobsen, 'A really worst case scenario. Rising seas from global warming may imperil coastal cities and low-lying islands', *Oceanus*, 32 (2), 37-43 (1989); U. Mikolajewicz, B.D. Santer & E. Maier-Reimer, 'Ocean response to greenhouse warming', *Nature, Lond.*, 345, 589-93 (1990).

39    T.H. Sibley & R.M. Strickland, 'Fisheries: some relations to climate change and marine environmental factors', in M.R. White, ed., *Characterisation of Information Requirements for Studies of $CO_2$ Effects*, United States Department of Energy, publ. DOE/ER-0236, 95-143 (1985).

40    G. Evelyn Hutchinson, 'Circular causal systems in ecology', *Annals of the New York Academy of Sciences*, 50, 221-46 (1948).

## NOTES TO FIGURES

41    After J.H. Steele, 'The message from the oceans', *Oceanus*, 32 (2), 5-9.

42    Data from G. Manley, 'Central England temperatures: monthly means from 1659 to 1973', *Quarterly Journal of the Royal Meteorological Society*, 100, 389-405 (1974); H.H. Lamb, *Climate. Present, Past and Future. 2. Climatic History and the Future* (Methuen, 1977) extended to 1987.

43    After M.C. McCracken, M.I Budyko, A.D. Hecht, A.D. & Y.A. Izrael, eds., *Prospects for Future Climate* (Lewis Publishers, Chelsea, Mich., 1990).

44    Data from H.H. Lamb, *Climate. Present, Past and Future. 2. Climatic History and the Future* (Methuen,1977), extended to 1989 by courtesy of C. K. Folland of the Hadley Centre for Climate Prediction and Research, Meteorological Office, Bracknell.

45    Original Plymouth data.

46    Data from City Environmental Health Officer, Plymouth.

47    H. H. Lamb, *Weather, Climate and Human Affairs* (Routledge, 1988).

48    Data from C.D. Keeling, *Atmospheric $CO_2$ Concentrations — Mauna Loa Observatory, Hawaii 1958-1986*, Environmental Sciences Divi-

sion, U.S. Department of Energy, publ. 2798, (1986), updated on disc to 1989.

49   After M.C. McCracken, M.I Budyko, A.D. Hecht, A.D. & Y.A. Izrael, eds, *Prospects for Future Climate* (Lewis Publishers, Chelsea, Mich., 1990).

50   After M.C. McCracken, M.I Budyko, A.D. Hecht, A.D. & Y.A. Izrael, eds, *Prospects for Future Climate* (Lewis Publishers, Chelsea, Mich., 1990).

51   After J.M. Barnola, D. Raynaud, Y.S. Korotkevich & C. Lorius, 'Vostok ice core provides 160,000-year record of atmospheric $CO_2$', *Nature, Lond.*, 329, 408-14.

52   After M.C. McCracken, M.I Budyko, A.D. Hecht, A.D. & Y.A. Izrael, eds, *Prospects for Future Climate* (Lewis Publishers, Chelsea, Mich., 1990).

53   Based on D. Rind, A. Rosenzweig & C. Rosenzweig, 'Modelling the future: a joint venture', *Nature, Lond.* 334, 483-6 (1988); M.E. Schlesinger & X-J. Jiang, ' Revised projection of future greenhouse warming', *Nature, Lond.* 350, 219-23 (1991).

54   Data from B. Bolin, ed., *Carbon Cycle Modelling* (J. Wiley, New York, 1981).

55   After M.C. McCracken, M.I Budyko, A.D. Hecht, A.D. & Y.A. Izrael, eds, *Prospects for Future Climate* (Lewis Publishers, Chelsea, Mich., 1990).

56   After J.L. Jacobsen, 'A really worst case scenario. Rising seas from global warming may imperil coastal cities and low-lying islands', *Oceanus*, 32 (2), 37-43 (1989).

# SAILING-SHIP SEAFARERS AND SEA CREATURES

## Michael Stammers

### Introduction

This paper concentrates on the nineteenth and early twentieth centuries and the seafarers of the merchant sailing ships of this period. It is divided into two main sections: the first examines the seafarer's relationship with the other creatures he encountered while at sea and the second some master mariners who contributed to the science of natural history. Sources are mainly printed and range from professional journals such as the *Nautical Magazine* (founded in 1832) to biographies and diaries which were published both as books and articles in enthusiasts' magazines such as *Sea Breezes* from the 1920s. I have sorted out from the mass of texts a number of 'benchmark' quotations which to me seemed to illustrate typical encounters, attitudes or beliefs. It must also be pointed out that most of the accounts were written by masters, officers and apprentices and not by the forecastle hands. The beliefs and attitudes of the latter are always reported second-hand through the eyes and ears of the upper deck or through the accounts of passengers. Certainly apprentices must have had a closer relationship with the hands than the officers. Apprentices worked alongside them and in some cases shared the same accommodation.

### Man and the Sea

Man's attitude to the sea has always been an ambivalent one from early times. It was the provider of great riches - a source of food, a means of communication and at the same time powerful, capricious and destructive. Its varied moods suggested a living creature or god to be worshipped and propitiated. Some peoples saw it as a sea creature - a sea serpent in some parts of the Pacific or in classical times - a human being, Oceanus or later Neptune. The latter survived into Christian times as the figure who governed the 'Crossing the Line' ceremony. By the period under consideration, he was no more than a figure of entertainment and superstition. But perhaps, there was unconscious thought that the ceremony ought to be carried out just in case - much as we touch wood - and there was a whole body of sailor's superstitions connected with animals and birds together with all kinds of mythical sea creatures ranging from the Kraken to mermaids. These have been exhaustively chronicled by the folklorists and they seem to have formed a body of belief which was traditional to seafaring. At the beginning of the nineteenth century before technical change had any real impact there is a case for believing in a continuity of oral tradition and belief that stretched back to the early seafarers. There was also the biblical teaching which made Man a unique creation of God who had given him command over the earth and its resources, including the creatures of the sea and the shore.

But the account of creation in Genesis came under increasing scrutiny in the first half of the nineteenth century. A literal seven-day creation estimated to have taken place 4,000 years earlier was contradicted by the study of fossil evidence which showed a time span of millions of years of development. This culminated with the publication of Darwin's *Origin of the Species* in 1859 which was based on his fieldwork on the scientific voyage of H.M.S. *Beagle* between 1831 and 1836. Darwin's conclusion that all life including Man had evolved by a process of 'survival of the fittest' caused an enormous debate which spread beyond the universities and the scholarly journals to the popular press. No member of the reading public including seafarers could fail to be aware of the controversy. Apart from challenging the biblical authority, it also lent scientific justification to utilitarian economic theory which proposed that interference (especially government interference through legislation) with the market and competition was bad and it bolstered the idea that European man, especially the British, were the summit of evolution and quite justified in taking over the lands and lives of less developed races - all of which had implications for seafarers. That God-fearing Welshman, Captain Robert Thomas, when writing to his daughters about the people of Tierra del Fuego in 1883, illustrates this:

> They are reckoned about the most miserable and lowest of the human race except the old natives of Australia who are in intellect very little superior to the monkey. But we must not forget that God made them as well as us and we ought to be thankful for a greater share of common sense ....[1]

An analysis of the mass of sailing ship voyage information in the sources already mentioned brings out certain common preoccupations. These include: joining the ship, departure, the character of the captain and to a lesser extent the officers, food, work on board, bad weather, accidents, landfall and life ashore including desertion. The most important environmental features were the weather and the sea, the tides, the currents, the storms, the calms and how the ship and her crew did or did not survive them. As well as the biographical accounts there was a large amount of technical literature on meteorology and hydrography. Maury's work in the 1840s and 1850s stimulated such studies which involved many working mariners making observations at sea. Incidentally, there is nothing on pollution of the sea. This present-day anxiety does not seem to have registered at all. Compared with work on board and the weather, sea creatures get little coverage and it is chiefly sharks, flying fish and albatrosses, or when there is a strange or unusual phenomenon. This may be because at times there was little or no wild life to record, and that such as there was of little interest unless the recorder happened to be interested in natural history and mainly that the work of the ship and the relationships between those on board were all absorbing along with the dominating factor of the weather. The task was to move the ship to deliver her cargo. Four hours on, four off, steering, navigating, setting sail, getting it in, reefing, tacking, wearing, changing sails, overhauling the rigging and its fixtures and fittings, maintaining the decks and the upper parts of the hull left little time for anyone on board to give little more than a passing glance to what was beyond. They were isolated in the sense that the freeboard of most vessels put them above the surface of the waves. But there were times when the sails did not have to be trimmed, when the weather was fine or set fair, or calm, which gave opportunities to observe and hunt the sea creatures around the ship. The steady trade winds was one time, another might be in the doldrums when the captain might get the ship's boats over the side, for exercise and entertainment. Another occasion would be while the ship was lying in a foreign port waiting to load or unload cargo.

**Animals on Board**

Many creatures travelled on and in the ship and all affected the crew. The teredo worm (*Teredo navalis*) indeed could destroy the very structure of the ship! This mollusc which was related to the cockle and the mussel was capable of boring long holes in ship's timbers. There was usually nothing on the outside of it to indicate their presence. The tropical species work faster (real damage became apparent in six weeks) than those in colder waters as the early European explorers found. Ship's timbers had to be protected by sheathing, layers of sacrificial wood, sheet lead and in the late eighteenth century, copper sheet or copper compounds, for example, muntz metal, a compound of copper and zinc. Aside from teredo and its many genera, there were gribble (*Limnaria lignorum*) that looked like a marine woodlouse, and barnacles and seaweeds which could attach themselves to the hull and slow the ship down. The latter problem was acute for the iron hull vessels from the mid-nineteenth century. They could not be sheathed in copper because of the resulting electrolytic action and anti-fouling paints and many were patented but few that were effective for long. The problem remains to this day.

On board, other creatures lived with and on men; hygiene was often not of high order, with personal parasites such as lice and fleas and bed bugs. Eric Newby, an English apprentice on the Finnish four-masted barque *Moshula* in 1938 recalled:

> I opened Captain Jutsum's book on knots and splices and out fell a big red bug, big because he had been feeding on me. I lifted the Donkey's breakfast. Underneath it, on the bunk boards was a small piece of canvas folded in half, it was full of them. I lifted the bed boards and the frames were seething.[2]

The problem was especially acute on ships carrying emigrants where officers often took dictatorial powers to bathe the offenders to prevent their spread to the rest of the passengers. Thus Dr Robert Skirving in 1883:

> I am sorry to have to admit that lice and bugs were a veritable plague. It was useless for angry viragoes with arms akimbo to shout taunts of lousiness to their next door bed-mates. Ah no! For all shared the attention of parasites of the best known brands. I solemnly declare     that when sweeping of the 'tween decks took place, I have looked into the line of dust and seen mobs of our little guests crawling about. I encouraged the men to have their hair cut very close; but with the women the days of bobbed hair had not yet arrived, and I fear the 'glory of womanhood' long locks, made fine converts for these little brothers of the poor and dirty.[3]

Rats and cockroaches were both common shipboard inhabitants; and some creatures were loaded with the cargo such as scorpions and spiders in tropical produce or flies hatched from maggots secreted in shipments of raw hides. Many ships carried domesticated animals : pigs and chickens for fresh meat and sometimes cows on passenger vessels for milk; and pets were usually the property of the master, such as cats and dogs or caged birds. Basil Lubbock mentions a canary in the half deck.[4] Cats were particularly popular because of

147

their rat catching and some sailors regarded the welfare of the cat as a matter of superstition. On the *Africaine* in 1836, there was an outcry when an cat disappeared overboard: 'I heard a man say he would hang a man for a glass of grog but would not drown a cat for a sovereign, for they consider it a certain forerunner to a storm.'[5] Cats or pets of any kind represented company and diversion. The master occupied a particularly lonely situation unless his wife or family accompanied him. Although the pet might have belonged to the master, it often became the pet of everybody on board, a mascot and a source of pleasure, a slight compensation for loss of family and children. So Captain Robert Thomas to his daughters in 1883:

> There, what good times a cat on board a ship gets to a shore cat. A cat or a dog are nursed like children and often all hands are watching and laughing at a cat playing, what a shoreman would never condescend to notice, but our pleasures are so few that we make the most of everything, however small.[6]

## Sea Creatures

Of the wild creatures, in the seas and skies surrounding the ship there was no such affection. Some were outright enemies, and others sources of food to supplement salt meat rations and sport. Sharks were feared; it was believed that all sharks would attack humans if they were unlucky enough to fall overboard. If sharks were sighted in fine weather, then shark fishing became the official ship's sport. The heavy hook and a swivel with chain trace either belonged to the master or the ship and had to be stout enough to withstand the power of the beast. Even a six footer had a lot of strength. 'A shark line could only be set with the Captain or the mate, chiefly because of the mess made on deck by blood and oil when the shark was caught.'[7] Seamen were always eager to catch one for four good reasons: the shark's tail nailed to the bowsprit brought fair winds, the death of the shark was an act of revenge and justice for the sailors it had probably eaten, its stomach was a lottery with rings, coins and other indigestible objects and also a source of excitement in the Doldrums. Other fish were caught for sport and eating, Lubbock recorded on September 8th 1889:

> There has been a great deal of fishing off the bowsprit... I had a try, but was not successful. You want to trawl your bait (a bit of white linen makes as good a bait as anything else along the water), jumping it occasionally... we had

*Plate 1.* Dolphins and flying fish were commonly encountered on a deep-sea passage. Original from Anon, *Great Shipwrecks* (London, Edinburgh and New York, c.1878), 77.

*Plate 2.* The wandering albatross, probably sketched by Richard Watt, a passenger on the *Young Australia*, in 1864. Whereabouts of original manuscript diary unknown.

*Plate* 3.   Catching cape pigeons in the Gulf of Penas on board the *Sunbeam*; seabird catching was a common sport for sailors and passengers. Original from Lady A. Brassey, *A Voyage in the Sunbeam* (London, 1891), 150.

*Plate 4.* Examining a 'haul' on board the *Challenger*. The dredge used on this warship in 1876 was similar to those employed by Captain Cawne Warren and other Liverpool-based mariner naturalists for collecting specimens. Original from F. Whymper, *The Sea: Its Stirring Story of Adventure, Peril and Heroism* (London, Paris and New York, c.1880), 1.

Loring's bonita for breakfast in the half deck. I don't think
any of these deep-water fish are much good eating, being
coarse and without much flavour but they are welcome on a
hungry 'lime juicer'.[8]

Flying fish, which often landed on board also made good eating. Most
sailors avoided eating shark for this was almost cannibalism. Every
shark was guilty of eating sailors. Sea creatures could also evoke a
sense of wonder and admiration among seafarers. The beauty of the
flying fish was admired, and birds, whales, dolphins, porpoises and
seals could all attract interest:

> On the look out one clear moonlight night, the ship scud-
> ding cheerfully along, I stood meditatively regarding the
> various constellations which studded the nocturnal
> heavens, indulging in ethereal cogitations, when I was sud-
> denly brought back to consciousness by something vibrant
> and glistening coming into contact with my breast and
> immediately a gusting sound on the deck discovered to me
> one of those tiny elves of the sea - a flying fish.[9]

At the same time, sailors would also try and harpoon such wonders.

The most beautiful, the most awe inspiring of all the wild creatures encountered on a sailing ship's ocean passage was the albatross
(*Diomeda exulans*) - an inhabitant of the southern hemisphere, sometimes with a wingspan of over 11 feet. It could soar behind a ship for
weeks with scarcely a movement of its wings. While many diarists
appreciated its beauty, this did not stop the sport of fishing for
albatross - nor did the superstition about the albatross popularised in
Coleridge's poem, the Ancient Mariner, have much currency:

> There is a general belief ashore that these birds are greatly
> respected  by sailormen and that a superstitious opinion
> exists amongst them that the destruction of an albatross
> brings dire calamity or at best head winds. I saw no evi-
> dence of this belief with the exception of one  or two of the
> oldest who sadly shook their heads and muttered
> solemnly when they saw the birds being hauled on board.[10]

Albatrosses were caught with a hook on a float with some meat as bait
tied to it, which was trailed astern. Catching or shooting birds was
considered good sport and an end in itself, although parts of the
albatross could be made into souvenirs such as tobacco pouches from

its large webbed feet and pipe stems from their bones. Their flesh was too oily to make good eating. The sailor's attitude corresponded with those of the landsman. Field sports on the land were considered an important and noble activity and the scale of the slaughter was even greater. The great 'battues' of the country estates saw thousands of game birds slaughtered in a day's shooting and in India and Africa, tigers, lions, elephants, buffalo, were despatched in large numbers by big game hunters. Landsmen enjoyed such sport while at sea; Richard Watt, a passenger on the *Young Australia* in 1864 reported:

> We have had several flocks of birds following the ship today including cape pigeons, whale birds and an albatross. Soon our 'crackshots' [both passengers and officers] were at the stern with rifles and fowling pieces and among them bagged three. The albatross, however, proved invulnerable and seemed to treat all pellets with contempt. Bait on a line was equally unsuccessful.[11]

It is interesting to note that some of the least of later apprentices' accounts do not approve of the albatross killing. Neil Campbell, for example, gives a nauseating account of the catching and deliberate maiming of an albatross by the mate who sawed off its beak and then threw it back in the sea to die slowly. The mate's justification was as an act of revenge, after having witnessed an apprentice who had fallen overboard being attacked by albatrosses.[12] Captain Walter Parker was almost another victim to the birds when the ship's boat capsized: 'Albatrosses, molly-hawk and cape pigeon circled over us, sweeping down within a few inches from our heads, hesitating and waiting to attack us as soon as we showed signs of giving in.'[13] So, to some sailors the albatross could be as much an object of hate as the shark.

**The Commercial Exploitation of Sea Creatures**

The killing of albatrosses and sharks had a minor impact compared with the results of the commercial exploitation of wild life by men in sailing ships - whaling, sealing, fishing. The quantities involved were huge. The main campaign can be traced back to the sixteenth century - the British, French, Basque and others found the great bonanza on the Grand Banks of Newfoundland and off the coasts of Canada - cod for salting, whales and other catches, while the offshore islands could supply birds and eggs as well as water for re-provisioning. The slaughter was huge too - and vulnerable species such as the great northern auk, the Arctic penguin, became extinct. The once plentiful whales of the Arctic had become increasingly scarce by the nineteenth century and new quarry was sought in the South Atlantic and Pacific. The process of extinction was much accelerated by the development of steam whale-catchers, harpoon guns and factory ships at the end of the nineteenth century. This overfishing, the near extinction or extinction of whole species of sea-going creatures, was a process that began in the sailing ship era; it was not just a twentieth-century phenomenon. The exploitation of nature had biblical sanction, and in many species the stocks seemed so abundant as to be inexhaustible. This attitude seems to have been universal and unquestioned among seafarers; and it was in a sense not undermined but reinforced by the theories of Darwin which put man at the top of the evolutionary ladder.

## Seagoing Natural Historians

To an extent, all seafarers were natural historians; birds were observed to give an indication of the nearness of land; whalers closely observed the habits of different species of whales (the Scoresbys actually wrote up their observations); and fishermen were experienced in their habitats of their catches. But systematic scientific observation started with the Royal Navy. Natural history combined with exploration, hydrography, navigational improvements, became a major area of scientific progress from the later eighteenth century. The success of Captain Cook's expeditions could be seen to mark the start of such a process. Other factors such as the development of scientifically-based classification systems developed by Linnaeus and others contributed. From the *Endeavour* voyage (starting in 1768) there were many official expeditions - and knowledge of marine and shore zoology and botany grew rapidly as a result. The ships were naval and those doing the collecting were usually specialist naturalists - for example Darwin on the Beagle in 1831 or Thomas Huxley in H.M.S. *Rattlesnake* in 1846, and there were also some highly competent amateurs among the ships' officers. As early as 1832, the *Nautical Magazine* was enthusiastically advocating naval and merchant navy officers making records of their voyages: 'There is probably no class of society which has more frequent opportunities of adding to the general stock of scientific knowledge than that composed of persons in the Royal and Mercantile navy'. In the same volume there are 'Hints for collecting specimens illustrative of zoology' and 'Directions for preserving plants in foreign countries' and 'Instructions for the collection of geological collections.'[14]

The *Nautical Magazine's* pleas do not seem to have been immediately taken up by the merchant service. This may have been because there was not time for such activities and also, those who did observe and record scientific and natural phenomena were more concerned with those with practical applications. In the late 1840s and '50s the *Nautical Magazine* contained a lot of correspondence recording positions of unknown offshore rocks, meteorological records, wind patterns, currents, navigation and experiments in 'great circle' routes, all of which had practical applications. It does become more mercantile in its readership from about 1850, but the natural history content is small and confined to curiosities. Nevertheless, there were collectors among seafarers especially among apprentices experiencing 'the wonders of the deep' for the first time. Many of their diaries and accounts mention collecting sargasso weed in bottles or preserving shark's jawbones, albatross heads or the wings of flying fishes, along with exotic shells from tropical ports. Likewise, the collecting at sea by merchant navy sailors was usually for souvenirs, for instance Sargasso weed in bottles. But there was no systematic collecting or observation. Many show some love and wonder for nature and a detailed knowledge and acute observing skills acquired over years at sea. Captain Robert Thomas described,

> the Mother Carey's chickens ... perhaps a very little larger than a swallow and a little darker colour and not quite so smart on the wind, but they follow a ship through all weathers warm and cold, rough and fine, and almost always on the wing. It's very seldom they are seen sitting on the water. They fly and skip close to the water astern of the  ship, just touching the water with their feet and picking up anything small as eatable but never rest to eat it but just jumping on and off the surface. Some time they are called the storm petrel, for they seem to be more at home in a gale than when it is fine weather. Old sailors [used] to be very superstitious and are yet to a great extent, and some of them actually believe that these petrels never go ashore but that they hatch their eggs under their wing.[15]

Captain Walter Parker was not only undecided as to whether he approved of catching albatrosses but actually did some tagging:

> In later years, I frequently let them go again, after attaching a small tin tally to one of the legs, with the date, lat. and long., and ship inscribed. I know other men have done the same thing but I never caught a bird that had been caught before, to my knowledge.[16]

In 1861, the systematic collection and study of natural history by seafarers took a step forward with the publication of a paper in the

*Proceedings of the Literary and Philosophical Society of Liverpool* by Dr. Cuthbert Collingwood, a Liverpool doctor. Collingwood had also presented this paper at the British Association Conference held at Manchester in 1861. It was entitled: On the Opportunities of Advancing Science enjoyed by the Mercantile Marine. Collingwood stated that Liverpool received 4,258 ships from foreign countries and colonies in a year (1857) and handled one-third of the commerce of England, and:

> How is it that such a vast staff of enterprising men, constantly sailing to all parts of the globe, do so little to add to our knowledge of the natural productions, which they, of all men are in the very best position to explore and best able to provide for the investigations of scientific naturalists at home.

Collingwood saw that mariners could gather specimens which could be brought back for study and cataloguing by the scientists. Nothing of this happened at the time: 'No accessions of importance are derived to our museums and collections from the labours of seafaring men. A piece of coral, a parrot, a shell or two or something which has received attention from its oddity is occasionally brought back to the sailor from the rich and interesting regions which he has visited'.[17] There was no system of recording information about the finds, such as position, nor knowledge of how to preserve them.

Collingwood directed his remarks to Captains. He accepted that apprentices and mates were likely to be too busy with ship duties, but:

> when he is entrusted with a command, the case is different. He is no longer a servant on board his vessel, but a master - his life of active employment is changed to one of comparative idleness; and it is well if the time thus left on his hands is not put to an evil use.[18]

He mentions cases of cruelty and intemperance. He also saw that shipowners might object to the encouragement of the study of natural history and the collection of specimens because it might cause masters to neglect their duties. This, said Collingwood, was a short-sighted narrow view, because the master did not have duties all the time; there were seasons of repose. Shipowners should be persuaded that this would be a positive benefit: 'a man with a hobby is always safer both at sea and on shore, than a thoroughly idle man.'[19] Collingwood urged the setting up of a committee to confer with Liverpool shipowners to set up a scheme which would be of mutual benefit - better masters and more natural history specimens. The paper was circulated to shipowners and a meeting held in the Town Hall to try and persuade local shipowners of the merits of the idea. This was chaired by a shipowners' representative, Thomas Miller Mackay, a major partner in the Black Ball Line, who took a positive interest in improving the training of seamen - and Captain Price of the Mercantile Marine Service Association, a professional organisation for master mariners established at Liverpool in 1859. The shipowners promised support and the MMSA was to prepare certificates for good work. The Literary and Philosophical Society for its part published a handbook on how to collect specimens with instructions on skinning and preservation, how to use dredges and trawls, microscopes, and keeping aquaria at sea. This was published in 1862 and copies were apparently circulated, while other learned societies were also contacted in the hope that they would set up similar schemes in other ports.[20] The Society also created a special type of associate member, limited to 25 at any one time, for master mariners who collected and studied natural history. Although it never exceeded more than 18 and petered out by 1900, nevertheless its members did make a useful contribution to the work of the natural scientists and this is attested by the stock books of Liverpool Museum. The most notable contributor was Captain Cawne Warren, elected in 1869 when master of the *Brilliant*. He collected material from then until his death in 1889. In 1882, for example, he returned from a voyage to Valparaiso with specimens of fish, kelp, algae, sponges, sea urchins, the contents of fishes' stomachs and barnacles. Some of them were new to science, for example, his 52 specimens of sponges and worms dredged up from the gulf of Manaar, which were identified by H. J. Carter and reported in the *Annals and Magazine of Natural History* in June 1880.

In conclusion, the average seafarer's relationship to the wild creatures of the sea and the sky remained much the same throughout the nineteenth and early twentieth centuries, and cannot be differentiated from the landsman. Marine creatures were a source of food and sport, but one cannot deny they were sources of wonder. Some apprentices' diaries indicate more than a passing interest, and in a few cases this may well have blossomed into substantial enquiry and knowledge when they became masters. A few elite Liverpool-based masters became involved in the observation, study and collection of natural history specimens and made an important contribution to marine biology in particular.

But, and most important of all, we should not forget the devastating effect of the commercial exploitation of wildlife - the whales, seals, walruses, fish, penguins, including the extinction of complete species - by seafarers, a process that continues to this day.

## NOTES

1   Aled Eames, 'Shipmaster, the life and letters of Captain Robert Thomas of Llandwrog and Liverpool, 1843-1903', *Caernarfon* (1980), 130.
2   Eric Newby, *The Last Grain Race* (London, 1956), 94.
3   Dr. R. Scott Skirving 'Recollections of a voyage on *Ellora*, Plymouth for Sydney, 1883', in *Medical Journal of Australia*, 27 June 1942, quoted in Don Charlwood, *The Long Farewell* (Ringwood, Victoria and London, 1981), 177.
4   Basil Lubbock, *Round the Horn before the Mast* (London, 1902), 153.
5   Mary Thomas's diary quoted in Penelope Hope, *The Voyage of the Africaine* (London, 1968), 65.
6   Eames, 'Shipmaster', 135.
7   Sir James Bisset and P.R. Stephenson, *Sail Ho! My Early Years at Sea* (London, 1958), 156.
8   Lubbock, *Round the Horn*, 114-5.
9   Journal of Jesse Neild, 1870-1, 54, private collection.
10  Neil Campbell, *Shadow and Sun* (London, 1947), 77.
11  Richard Watt, 'Second cabin passage', *Sea Breezes*, 21 (1956), 408.
12  Campbell, *Shadow and Sun*, 74-6.
13  Captain Walter H. Parker, *Leaves from an Unwritten Log-Book* (London, c.1935), 46.
14  *Nautical Magazine*, I (1832), 180, 303-7, 575, 578.
15  Eames, 'Shipmaster', 135.
16  Parker, *Leaves*, 20-1.
17  Cuthbert Collingwood in *Proceedings of the Literary and Philosophical Society of Liverpool*, 51st Session (1861-2) XV1, (Liverpool, 1862), 47-8.
18  *Ibid.*, 49.
19  *Ibid.*, 51.
20  The handbook was also published as appendix II of the Society's proceedings of 1862 under the title of 'Suggestions offered on the part of the Literary and Philosophical Society of Liverpool to members of the Mercantile Marine, who may be desirous of using the advantages they enjoy for the promotion of Science, in furtherance of Zoology'.

# DEVON SEASIDE TOURISM AND THE COASTAL ENVIRONMENT, 1750-1900

## John F. Travis

Nowhere was the destructive nature of the relationship between early tourism and the coastal environment more clearly demonstrated than at the Devon seaside. In many ways the relationship resembled an ill-fated romance; with the early tourist being cast in the role of an admirer unable to prevent himself defiling the attractive virgin environment he had come to pay suit to. The early Devon resorts had relatively few commercial attractions; it was the natural attractions of the sea and shore, the pure maritime air and the fine coastal landscapes which enabled them to win the patronage of wealthy holiday-makers. Yet sadly these priceless natural assets were steadily eroded, as the trickle of visitors became a steady stream and as the holiday resorts began to expand.[1] This paper sets out to identify those aspects of the Devon coastal environment which attracted visitors. It will show that in every case they were severely damaged by the growth of the holiday industry.

It was a change in medical fashion which first encouraged members of the leisured class to visit the English coast. Since the seventeenth century leading members of the medical profession had been advising their patients to 'take the waters' at the inland spas. Rich gentlefolk had congregated at these early watering places, some seeking an antidote for intemperate living or relief from a chronic complaint, others hoping that fashionable company and new forms of entertainment would deliver their minds from the malaise of boredom. As physicians sought new mineral waters to recommend to their wealthy clients, they at last began to realise that the sea offered a limitless supply of water, even richer in salts than that at the spas.[2]

In 1750 Dr Richard Russell published the first edition of his *Dissertation on the Use of Sea Water in the Diseases of the Glands*, which called public attention to the sea as a potential source of health. This Sussex doctor prescribed sea bathing not as a recreation, but as the taking of a medicated bath. He went one step further, for he also advised invalids to drink sea water. The thought of gulping down glasses of sea water may seem revolting to our generation, but it has to be remembered that at the spas the gentry were accustomed not only

160

to bathe, but also to imbibe large quantities of the nauseous spring waters. The success of Russell's book encouraged many other doctors to write books extolling the virtues of sea water taken both internally and externally.[3]   Sea-water treatments became fashionable and wealthy people began to seek both health and pleasure at the seaside.

Many of the early visitors to the Devon coast were lured there by the publicity being given to the health-giving properties of sea water. As early as the 1750s there were reports that some gentlefolk and merchants from Exeter were frequenting the nearby  fishing ports of Exmouth and Teignmouth in the summer months because the sea water there had been 'found beneficial'.  These two coastal villages became the first seaside resorts on the Devon coast, accommodating visitors arriving there 'both for health and recreation'.[4]  Then in 1771 Dr Downman, a celebrated Exeter doctor, published a long didactic poem entitled *Infancy* in which he suggested that invalids should visit Dawlish, where they would find that the 'pure encircling waves' would 'bestow a second happier birth from weakness into health'.[5] His recommendation was sufficient to cause the little coastal village to develop as another small health resort.[6]  In the same year there was the first report of a few strangers visiting Ilfracombe on the north Devon coast 'for the benefit of the ... salt water'.[7]  The lure of renewed health was obviously very strong, for the track ways they had to use on their way to this fishing village were in an appalling condition. Sidmouth likewise began to attract those seeking sea-water cures.  In 1776 it was noted that the fishing village was playing host to 'company resorting hither for the benefit of bathing and drinking the waters'.[8]

Quite remarkable claims were made for the healing properties of the sea water on the Devon coast.  In 1762 the *Royal Magazine* advised its readers that at Teignmouth:

> For the sake of drinking that fashionable purging draught, sea water, and bathing ... numbers of people from all parts resort here in the summer season, and cripples frequently recover the use of their limbs, hysterical ladies their spirits and even the lepers are cleansed.[9]

In 1789 the  *Exeter Flying Post* stated that at Teignmouth: 'Bathing in the sea and drinking the salt water have in many cases been attended with great success'.  The newspaper cited the case of an elderly clergyman who, after losing the use of both hands and legs, had visited Teignmouth in the hope of a cure:

> By bathing a few weeks, and drinking the sea water every other morning, he recovered the use of his limbs so far as to

be able to walk with crutches ... and before he left
Teignmouth was entirely restored.[10]

Pure sea water was obviously essential if the visitors hoped to
obtain improved health by drinking it as well as bathing in it. So it is
hardly surprising that the Devon seaside resorts boasted of the purity
of their marine 'elixir of life'. Sidmouth, for example, in 1791 claimed
that it had sea water 'in great perfection',[11] while Exmouth in the same
year announced that visitors to its shores 'dipped up some of the
purest sea water that can be drunk'.[12]

Unfortunately, once the tourist trade began to develop, sewage
from the growing resorts inevitably polluted the sea near the points of
discharge. Even in the late eighteenth century some sewage had been
allowed to run down pipes into the sea, but in those early days there
were no proper sewerage systems and most of the human excreta had
been collected in cesspits. There was soon a desperate need for better
disposal systems to cope with the sewage produced by the growing
visitor and resident populations. By the mid-nineteenth century all
the principal Devon seaside resorts were adopting the cheap and easy
solution to the problem; running sewers across the beaches so that all
their raw sewage could be discharged into the sea.[13]

So the sea water that holiday-makers regarded as a source of
good health was actually being contaminated by their visits. At
Ilfracombe, for example, visitors bathed in the vicinity of the harbour,
but that same harbour in 1850 was described as a 'large open cesspit'.[14]
In the same year Torquay claimed that its visitors had the advantage
of an 'illimitable ocean bath of spotless purity',[15] but the reality was
very different. In fact Torquay's headlong expansion had created
serious coastal pollution and at low tide the stench was overpowering.
The problem was caused by three separate sewers spewing their contents straight across the beach, so that both the shoreline and the sea
were badly contaminated.[16]

In the late-nineteenth century increasing complaints of foul sea
water and polluted beaches obliged the leading Devon resorts to
embark on ambitious schemes to extend sewer outfalls below the high
water mark to points where tidal currents could carry away the
noxious waste. It was Torquay that led the way, in 1878 completing
an ambitious £65,000 project which involved boring long tunnels,
intercepting old sewers and laying a new main to an outlet north of
Hope Nose, where it was hoped the tide would sweep the effluent
well clear of the main beaches.[17] This major engineering scheme
attracted considerable publicity and focused local attention on the
need to carry sewage further out to sea. By the end of the century all
the leading Devon resorts had constructed new sewer outfalls to

*Plate 1.* Sidmouth, c.1829: as resorts expanded so smoke pollution inevitably increased. Reproduced by kind permission of the Devon and Exeter Institution from a steel engraving entitled *The York Hotel and Library, Sidmouth,* by P. Heath after W.H. Bartlett.

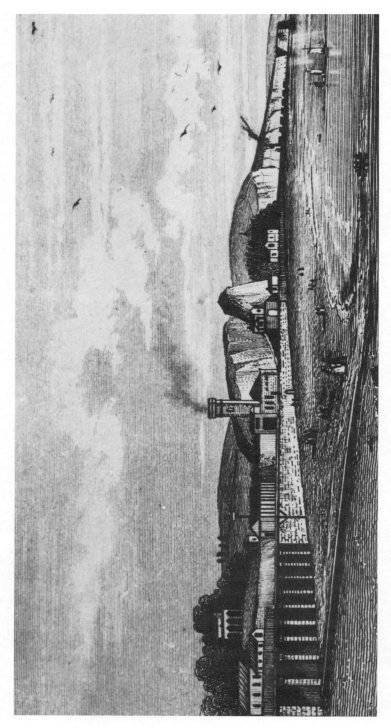

*Plate 2.* Dawlish, c.1848: smoke from the pumping station and an approaching train drifts across the beach. Reproduced by kind permission of Torquay Museum from a steel engraving entitled *Station and Engine Houses, Dawlish,* by J. Newman and Co.

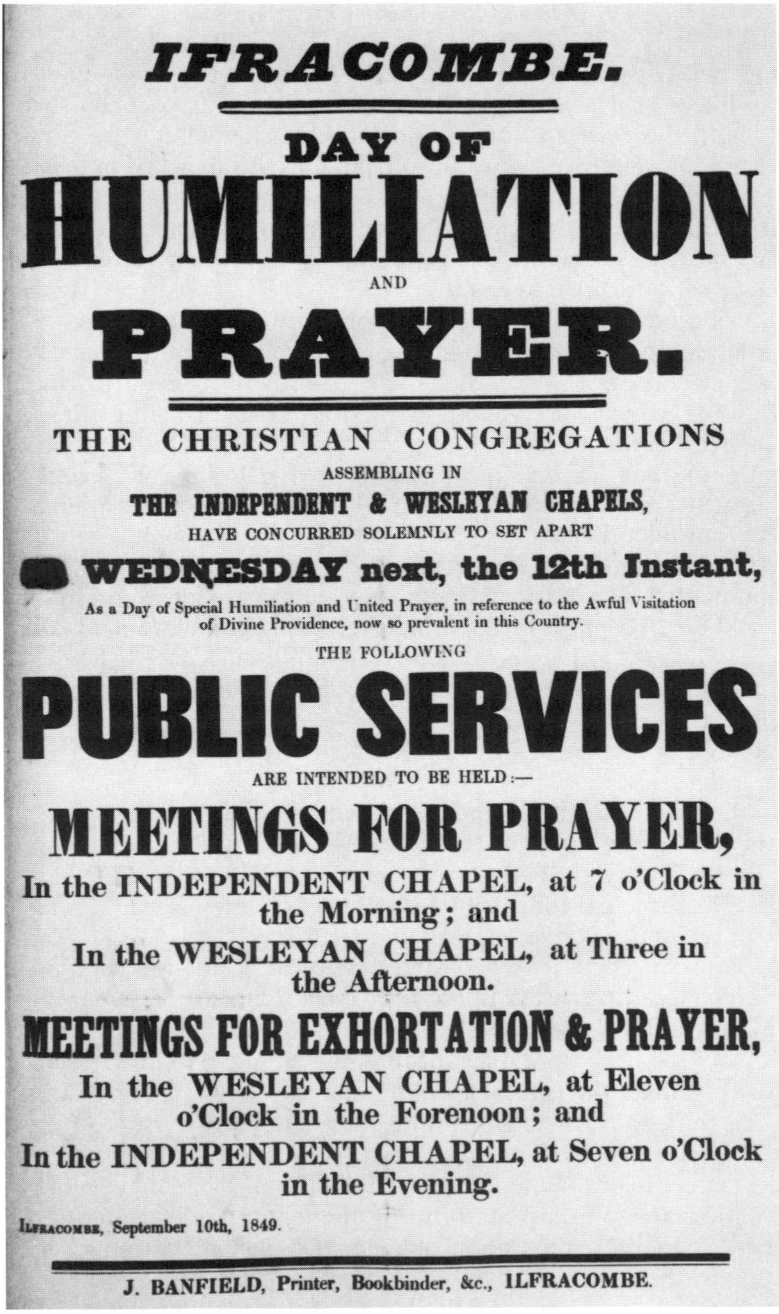

IFRACOMBE.

DAY OF

HUMILIATION

AND

PRAYER.

THE CHRISTIAN CONGREGATIONS

ASSEMBLING IN

THE INDEPENDENT & WESLEYAN CHAPELS,

HAVE CONCURRED SOLEMNLY TO SET APART

WEDNESDAY next, the 12th Instant,

As a Day of Special Humiliation and United Prayer, in reference to the Awful Visitation of Divine Providence, now so prevalent in this Country.

THE FOLLOWING

PUBLIC SERVICES

ARE INTENDED TO BE HELD:—

MEETINGS FOR PRAYER,

In the INDEPENDENT CHAPEL, at 7 o'Clock in the Morning; and

In the WESLEYAN CHAPEL, at Three in the Afternoon.

MEETINGS FOR EXHORTATION & PRAYER,

In the WESLEYAN CHAPEL, at Eleven o'Clock in the Forenoon; and

In the INDEPENDENT CHAPEL, at Seven o'Clock in the Evening.

ILFRACOMBE, September 10th, 1849.

J. BANFIELD, Printer, Bookbinder, &c., ILFRACOMBE.

*Plate 3.* Poster produced at Ilfracombe during the cholera epidemic of 1849. Reproduced by kind permission of Ilfracombe Museum.

points below the low-tide mark, in the hope that the tide would carry away the offensive matter.[18] Their efforts were only partially success-ful. Both the Bristol Channel and the English Channel are confined seas and nothing could prevent some of the evils expelled at one point from washing up further along the coast. Sadly then the increase in tourism had meant that the sea water, which had long been promoted for its purity and therapeutic properties, was badly polluted and was in fact a serious threat to health.

Coupled with the preoccupation with sea-water cures was a parallel interest in the health-giving qualities of the pure sea air found on the Devon coast. As early as 1759 Andrew Brice reported that the citizens of Exeter were visiting Exmouth 'for the benefit of fine fresh air' as well as for the sea bathing, and Teignmouth, 'the air being very wholesome here especially in summer, wheretofore 'tis visited both for health and recreation'.[19] In 1771 Dr Downman declared in his poem that inland cities like Exeter had an 'unwholesome atmosphere, gravid with seeds of latent sickness' and that they were smoke-contaminated by a 'darkening plume of poison and death prolific'. He advised his patients to escape from the 'crowded town ... that court of death where every gale is tainted with pollution' to Dawlish where he claimed that the 'refreshing breeze' would help to restore them to full health.[20] In the same way early visitors to both Sidmouth and Ilfracombe went there seeking the 'benefit of the sea air' as well as that of the sea water.[21]

National interest in the therapeutic properties of sea air was stimulated by the work of Dr Ingenhousz, who in 1780 gave a paper to the Royal Society entitled *On the Degree of Salubrity of the Common Air at Sea*. His principal conclusion was: 'Air at sea and close to it is in gen-eral purer and fitter for animal life than the air on land'. This he argued was a good reason why the medical profession should send 'patients labouring under consumptive disorders to the sea or at least to places situated close to the sea'.[22] This paper had an important effect on medical thinking, encouraging a growing interest in the nature of climates at coastal locations and their influence on health.

At the end of the eighteenth century attention began to focus on the temperature as well as the purity of the Devon maritime air. Many delicate people of rank and fortune had been in the habit of wintering in Mediterranean France, but during the long wars that followed the French Revolution they were obliged to look for alternative winter havens in their own country. The climate of the south Devon coast then began to receive considerable publicity as the nearest English equivalent to the south of France. As early as 1791 Dr Jebb, physician to George III, announced that he would 'aver the pureness and salubrity of the air at Exmouth equal to the south of France'.[23] By 1803 a national guide could state: 'The mild and genial softness of the air on the south coast of Devon is generally esteemed equally salutary with that of Montpellier or Nice and is frequently prescribed for pulmonic disorders and declines'.[24] Exmouth, Teignmouth, Sidmouth and the infant resort of Torquay became the first coastal watering places in England to enjoy the advantages of a second season in winter. The south Devon coast was thought to offer a happy fusion of exceptionally mild winters and pure sea air, as an 1804 guidebook made clear:

> It presents its bosom only to the southern ray, and to the southern zephyr, and, fanned by the pure breeze of the ocean alone, must of course be well-calculated to redress the injury which filthy cities, crowded rooms and mephitic vapours entail upon mankind.[25]

The inhabitants of the overcrowded towns and cities of England hoped that the pure air and genial temperatures on the Devon coast would protect them from illness or restore them to full health. In an age of rampant disease, tuberculosis was feared most, for it was a merciless killer. There was no real remedy, but at the end of the eighteenth century eminent members of the medical profession began to send consumptives to the south Devon seaside where they believed that the maritime climate offered the prospect of a cure. The mild, pure air was soon being promoted as an almost universal panacea. One local publication in 1805 claimed that on this favoured coast the invalid could 'inhale those breezes which so frequently suspend the ravage of disease, pour fresh oil into the lap of life and send him back a renovated being'.[26]

Yet the development of the tourist industry soon began to threaten the purity of the sea air, just as it did the purity of the sea water. Holiday-makers had escaped from the atmospheric pollution of their inland towns hoping to find a healthier environment on the Devon coast. But as the Devon seaside resorts expanded in the nineteenth century, so there too more and more chimneys belched out more and more smoke from more and more coal fires. While the problems of atmospheric pollution were nothing like as serious as in many of the large industrial towns of England, the quality of the air at the seaside resorts was steadily deteriorating and was certainly nowhere near as pure as their publicity still pretended. Smoke pollution was obviously worst in the winter when there was a need for more coal fires, yet this was just the season when the majority of consumptives and victims of other lung complaints arrived on the south Devon coast, hoping to benefit from breathing pure maritime air.

Very few of the inhabitants of the resorts seemed aware of the threat to one of their principal natural assets. Octavian Blewitt, the author of an 1832 guide to Torquay, was a lone voice sounding the alarm. He pointed out that the boom in Torquay's tourist trade was bound to lead to more atmospheric pollution and could jeopardize its reputation as a health resort specialising in the treatment of lung complaints:

> The greater the increase of inhabitants ... and the nearer the place approximates to the character of a town, the more serious will be the injury to that climate which is now the source of its prosperity.[27]

When the rail network began to extend towards the Devon coast, the resorts were obsessed with the need to obtain a rail link which would make them accessible to far more visitors, and ignored the fact that the railways would inevitably give rise to more smoke. Perhaps small resorts expecting to be served by only a few trains a day could be excused their lack of concern, but, even at those resorts where there was the prospect of quite heavy traffic, the need to obtain a rail link seemed to override all other considerations and the inhabitants seemed unconcerned about the prospect of increased atmospheric pollution.

The case of Dawlish and Teignmouth illustrates the low priority given to preventing smoke pollution. In 1846 these two resorts became the first in Devon to be reached by the railway.[28] The South Devon Railway which served them was originally intended to be an atmospheric railway. Pumping-houses with huge chimneys were needed every few miles to create a vacuum in a pipe laid between the rails, which would draw along the trains. If the system had proved successful there would have been no need for locomotives to pull the trains. Yet it is significant that there was virtually no opposition when the railway company announced that it would be erecting pumping houses right in the centre of Dawlish and Teignmouth, instead of building them in open country where their smoke would cause less damage to the air the residents and visitors breathed.[29] Victorians were much more interested in technological progress and financial profit than in environmental protection.

There was yet another way in which the growth of the tourist industry posed a threat to the healthy environment the visitors expected to find on the Devon coast. The holiday-makers believed that they were escaping from insanitary, germ-ridden inland towns to salubrious seaside locations. It is true that when the Devon seaside resorts were in their infancy they could claim relative freedom from many of the health hazards besetting many rapidly growing urban areas in other parts of the country. They had grown up in relatively sparsely-populated rural areas so water could be easily obtained from unpolluted sources and the comparatively small quantities of sewage could be disposed of without great risks to health. Yet once the Devon watering places began to expand, they were faced with problems similar to those their visitors thought that they had left behind. The resident population at least doubled at all the major Devon resorts between 1801 and 1851 and at Torquay it increased more than 13 times.[30] The situation was made worse by the increase in the number of holiday-makers, for they placed additional strains on the totally inadequate systems of water supply and sewage disposal.

These problems finally led to serious epidemics at some of the Devon watering places, finally destroying the myth that the Devon coast was immune from the ravages of inland disease. There were, for example, major cholera epidemics at both Torquay and Ilfracombe in the late summer of 1849, which brought about many deaths. Panic set in and the visitors fled in terror. At Torquay it was stated: 'The part of the town where the deaths have occurred is almost wholly deserted',[31] while at Ilfracombe it was reported: 'The visitors have deserted the town causing a loss computed at not less than eight to ten thousand pounds'.[32] At Westward Ho! in the summer of 1875 a mystery epidemic, attributed by the Medical Officer to the effects of 'putrefying sewage' on the drinking water, left two children dead, many other people seriously ill and the tourist trade devastated.[33] Typhoid also visited several Devon resorts. At Ilfracombe, for instance, this highly infectious fever swept the prestigious Ilfracombe Hotel in 1878,[34] while at Lynmouth it endangered lives and livings in 1882 and again in 1884.[35] Reputations as health resorts were placed in jeopardy by such epidemics, so in the second half of the nineteenth century all the major coastal watering places felt obliged to embark on major public-health improvements.[36]

Splendid scenery also attracted many tourists to Devon, but once again the growth of the holiday industry gravely damaged one of the coast's principal natural attractions. The end of the eighteenth saw the development of the cult of the Picturesque. The Devon coast was just one of the remote areas of Britain that began to attract those seeking scenes that could delight the eye and inspire the artist's brush.[37] In the nineteenth century Devon's fine coastal landscapes drew increasing numbers of tourists, but the houses and hotels built to accommodate them began to obliterate some of the natural beauty surrounding the seaside resorts. It was only in the later part of the nineteenth century that a few voices were raised in protest. In 1873, for example, a letter appeared in the *Ilfracombe Chronicle* complaining about the

'destroyers' who were building large numbers of villas to accommodate holiday-makers, but in so doing were blotting out the fine
scenery the visitors came to see:

> Is it not a crime, under cover of improvement, to disfigure
> the natural beauty of the place? Ilfracombe of old, the resort
> of artists, tourists and lovers of anything beautiful, will ere
> long be spoilt.[38]

In the second half of the nineteenth century some discerning
tourists began to seek out smaller seaside resorts where they hoped to
find relatively unspoilt marine views. One visitor to Lynton and Lynmouth in 1886 rejoiced that the magnificent scenery surrounding the
resort was still unscathed, but warned the inhabitants to be wary of
'the speculative builder with his levelling mania, flattening, smoo-
thing, and parading all around, raising stucco terraces and marine
crescents, hideous to look upon, turning each seaside resort into one
monotonous uniformity of common-place'.[39] Only two years later his
worst fears were realised. The tiny resort's unique scenery was
threatened by the start of several building projects to provide more
tourist accommodation. One observer commented: 'The hand of man
is doing its usual fatal work on one of the loveliest spots our country
had to boast. Flaring notices everywhere proclaim the fact that building sites are procurable through the usual channels, this estate and the
other has been laid out'.[40]

Even the remotest stretches of the Devon seaboard were not
immune from the spoiling works of the developer. In the 1860s the
success of Charles Kingsley's book *Westward Ho!* led to the creation of
a completely new resort bearing that name on the virgin shores of
Northam Burrows. The company responsible for the development
hoped that more tourists would arrive in north Devon to see for themselves the coastal landscapes so affectionately portrayed in this grip-
ping epic.[41] Charles Kingsley was far from pleased to hear that he was
indirectly responsible for the violation of a previously unsullied coastal region. In a letter written in 1864 to one of the directors of the new
company, Kingsley expressed his bitter regret at the construction of a
holiday complex on one of his favourite sections of the Devon coast:

> How goes on the Northam Burrows scheme for spoiling the
> beautiful place with hotels and villas? I suppose it must be,
> but you will frighten away all the sea-pies and defile the
> Pebble Ridge with chicken bones and sandwich scraps.[42]

Not even the natural habitat of the shore below the high-tide
mark was safe from the effects of tourism. The varied rock formations

and warm sheltered waters nurtured a rich diversity of animal and vegetable life unequalled perhaps in any other part of Great Britain. It was on the Devon coast that Philip Gosse, an eminent Victorian naturalist, in the 1850s conducted much of his pioneer research on the living organisms found in the tidal rock pools.[43] The spate of books that he and his followers published set in motion the invasion of the Devon coast by serious-minded Victorians eager to combine the advantages of an intellectually-improving activity with the pleasures of a seaside vacation. Marine biology soon became a consuming craze at the Devon resorts. Visitors scrambled over the rocks, searching for rare specimens to capture and take back to aquariums in their lodgings. Ilfracombe was the principal centre of this new craze and from there one visitor wrote in 1863:

> Zoophytes are quite an article of commerce here and a brisk trade is driven in naturalists' implements - knives for slicing them off the rocks, nets like dumpling-skimmers for catching the more erratic cephalopods and pots and pans with every fancy for the aquarium.[44]

Sadly this sudden surge of interest in the wonders of the Devon shoreline had an unforeseen effect. An army of collectors arrived on the Devon coast. They armed themselves with crowbars, chisels, knives and nets, and set to work to strip the Devon rock pools bare. By 1865 Gosse was expressing dismay at the pillage:

> Since the opening of the sea-science to the million, such has been the invasion of the shore by crinoline and collecting jars, that you may search all the likely and promising rocks within reach of Torquay, which a few years ago were like gardens with full-blossomed anemones and anthems, and come home with an empty jar and an aching heart, all being now swept as clear as the palm of your hand.[45]

By the end of the century it was thought that the Devon tidal pools had suffered irreparable damage. Edmund Gosse wrote in sad terms of the destruction of these submarine gardens brought about by the success of his father's books:

> They exist no longer, they are all profaned, and emptied and vulgarized. An army of collectors has passed over them and ravaged every corner of them. The fairy paradise has been violated ... crushed under the rough paw of well-meaning, idle-minded curiosity.[46]

Philip Gosse had encouraged amateur naturalists to visit the Devon seaside and these visitors had wreaked havoc on the very rock pools that they had come to marvel at. Just as Charles Kingsley had been distressed by the unforeseen damage to the coastal environment caused by his book *Westward Ho!*, so Gosse was equally upset at the destruction that he had unintentionally helped to bring about. Edmund Gosse commented:

> That my father, himself so reverent, so conservative, had by the popularity of his books acquired the direct responsibility for a calamity that he had never anticipated, became clear enough to himself before many years had passed, and cost him great chagrin.[47]

The natural habitat of the shoreline pools was also damaged by the increasing quantities of raw sewage being discharged into the sea from the growing resorts. Charles Kingsley frequently holidayed on the Devon coast in the 1850s and 1860s,[48] at a time when many of the expanding resorts were constructing new sewerage systems with outfall discharging untreated sewage into the sea. He was a keen marine biologist, and well ahead of his time, for he was one of the very few people who seemed aware of the damage being done to the coastal environment. When in 1863 he published his famous children's classic *The Water Babies*, he used it to warn his readers of the harm that raw sewage could cause to the plants and animals living in the sea:

> Only when men are wasteful and dirty, and let sewers run into the sea instead of putting the stuff upon the fields like thrifty reasonable souls ... there the water babies will not come ... but leave the sea anemones and the crabs to clear away everything, till the good tidy sea has covered up all the dirt in soft mud and clean sand, where the water babies can plant live cockles and whelks and razor-shells and sea-cucumbers and golden combs, and make a pretty, live garden again, after man's dirt is cleared away.[49]

Another of the attractions of the Devon coastal environment was its quiet tranquillity, but this too was threatened as visitor numbers grew. In their early days the Devon seaside resorts were regarded as havens of peace far removed from the bustle and glamour of busy inland towns and large commercialized seaside resorts. In 1803 Torquay could accurately be described as appealing mainly to 'the lover of simple nature who can dispense with crowded assemblies, gaming tables and a train of luxurious refinements',[50] while in 1828 Ilfracombe could fairly be promoted as having 'retired but soul-enhancing charms'.[51] Yet, as the Devon coastal watering places began to expand, so visitors found it increasingly difficult to escape noise.

Music was a particular problem. In the second half of the nineteenth century the larger Devon resorts all rang with band music, usually performed by professional groups of German, Italian or Hungarian musicians. Normally comprising between 10 and 15 performers, they depended for their income on the money that their collector could badger out of their audiences and from local subscribers. They were known as the 'season band' because they returned for the summer season each year 'as regular as the swallows'. While some residents and holiday-makers enjoyed their musical renderings, others found them a source of great annoyance. At Ilfracombe, for example, there was in 1863 a particularly vitriolic press campaign against the 'execrable German band who have hitherto almost scared away visitors by their discordant noise and unintelligible jargon'.[52] It was certainly difficult for visitors to escape these season bands, for at all of the principal resorts they performed each day on the promenades, piers, squares and other public places. Not infrequently two rival bands tried to drown each other out as they battled for the financial support of the wealthy sojourners. At Ilfracombe in 1876 an English and a German band played in opposition to each other for most of the summer. Residents and visitors were outraged:

> There is on the Parade nightly a terrible mixture of sounds which may be individually sweet, but in their compound state are abominable. The old and the new season bands, in close proximity, play against each other might and main. The effect on the ordinary ear is painful, but to the practical and cultivated musician it is simply unbearable.[53]

For most of the nineteenth century the Devon seaside resorts had at least been able to pride themselves that their beaches were relatively free from the noise of musical entertainers. Elsewhere on the English coast nigger minstrels might strum and croon, but at the select Devon seaside resorts such discordant strains were frowned upon. As late as 1898 one guidebook to the south Devon resorts could warn:

> There are no niggers on the sands or other garish pleasures, and holiday-makers who love the flesh-pots of Margate and Ramsgate had better keep away from the south Devon watering places.[54]

In fact this was no longer true. By the mid-1890s the Devon resorts were at last being invaded by groups of black and white minstrels. Within a few years these burnt-cork negroes had new rivals on the Devon beaches, for troupes of pierrots, with their faces as white as oxide of zinc could make them, were also adding to the musical clamour.[55]

By the turn of the century it was not even possible to escape the noise in streets away from the beach. In 1900 the *Torquay Times* complained:

> Will nothing be done to control the fiendish pandemonium which prevails in Torquay. Year by year the visitors complain bitterly of the organ grinders, the nondescript instrument torturers, the concertina players, and the whistling fiends who are set loose to torture the nerves of health-seeking invalids.[56]

It was becoming increasingly difficult to enjoy a quiet holiday at any of the principal Devon resorts.

In the late nineteenth century the peaceful atmosphere of the Devon resorts was also threatened by an influx of working-class excursionists. Rising real wages and improved transport facilities were making it possible for increasing numbers of poorer people from inland urban areas to go on day trips to the seaside. By the 1890s most of the leading Devon resorts were having to play host to hundreds of rail excursionists at weekends in the summer months, while on fine bank holidays Dawlish, Teignmouth, Seaton and Ilfracombe might expect up to 2,000 rail excursionists, Torquay over 3,000 and Exmouth, because of its proximity to Exeter, up to 6,000.[57] These numbers were small in comparison with the tens of thousands of trippers that were flooding into Blackpool, Margate and other large resorts within easy reach of the major urban areas.[58] But these excursionists could and often did disrupt the sedate atmosphere of the Devon resorts for the few hours that they were present. On August Bank Holiday 1887, for example, one observer noted that 'the ordinary visitors at Teignmouth were nearly banished by the crowd which surged upon the promenade, the beach, the machines, the river; and ate and smoked on the common and danced on the pier'.[59] The air of peaceful tranquillity, which for many years had attracted visitors to the Devon coast, was being seriously affected by the continuing development of the tourist industry.

Ilfracombe was most at risk from tripper incursions because it was easily reached by excursion steamers. From the 1840s onwards the resort had to entertain large groups of coal miners and metal workers arriving on day trips from south Wales. As early as June 1849 the local correspondent of the *North Devon Journal* reported that 'over 400 of the human swarm of Swansea' had invaded Ilfracombe and had disturbed the peace by racing donkeys through the streets.[60] As the number of steamer excursions increased so did the complaints. In August 1864, for example, the *North Devon Journal* criticised the behaviour of a party of 750 trippers from Port Talbot who had rampaged through the resort: 'The great majority of them, both men and women, became drunk before leaving and by their noise and violence created quite a stir'.[61] By the 1890s the problems were much worse, for sometimes in a single day as many as seven excursion steamers discharged passengers at the harbour. In 1896 the *Ilfracombe Chronicle* voiced the growing public concern that 'on many occasions drunkenness has caused unruly conduct in our town'.[62] These Welsh trippers were ravaging the peaceful atmosphere which had long been one of Ilfracombe's principal attractions.

The social environment of the Devon coast was also threatened by the commercialisation that tourism brought in its train. As sleepy coastal villages were transformed into thriving seaside resorts, so great changes were wrought in the lives of the local inhabitants and these changes were not always for the best. The case of Lynton and Lynmouth illustrates the point. Early visitors to the twin villages had been captivated by the appealing innocence of country folk living close to nature and seemingly reflecting the virtues of a less avaricious age. At the beginning of the nineteenth century one traveller had envied the local peasantry's remoteness from the 'bustle of the world and the cares of crowded life',[63] while another remarked that 'their constant diligence and cheerful character claim the respect of the higher classes'.[64] But, as the tourist trade developed, the simple dignity of an earlier era was corroded by the grasping greed of a new commercialism, and before long the visitor found himself being accosted by landladies, boatmen, guides and donkey boys, all touting for trade. The hotels also competed aggressively for custom. In 1851 one guidebook warned the intending visitor:

> At Lynton telescopes are employed at the rival houses for the prompt discovery of the approaching traveller. He had better therefore determine beforehand on his inn, or he will become a bone of convention to a triad of postboys, who wait with additional horses at the bottom of the hill to drag the coach to its destination.[65]

As elsewhere on the Devon coast, the development of the tourist industry brought considerable financial benefits to the seaside communities, but only at considerable cost to the apparent simplicity of life which early tourists had found so enchanting.

We have seen that by a strange irony the development of the tourist industry severely marred those features of the coastal environment which attracted visitors to the Devon seaside. Those involved in the holiday industry often seemed to be concerned only with immediate gain and frequently ignored the damage being done to the natural and social environment which was their principal asset. By the end of the nineteenth century it was clear that some holiday-makers were beginning to avoid the larger resorts, which had suffered most environmental damage, and instead were seeking unspoilt scenery, healthy conditions and solitude on remote stretches of the Devon coast. This led to the emergence of minor resorts at previously undeveloped locations. On the south coast, for example, holiday accommodation was being built at the coastal hamlets of Hope, Bantham and Torcross, while on the north coast Woolacombe, Croyde Bay and Saunton Sands were gearing up to play host to growing numbers of holiday-makers. Sadly, the arrival of these visitors at these isolated places once again triggered off the same destructive processes which had already inflicted lasting damage on the coastal environment at the established resorts.

## NOTES

1    For a full account of the early development of the Devon seaside resorts see, John F. Travis, *The Rise of the Devon Seaside Resorts, 1750-1900* (University of Exeter Press, 1993).

2    J.A.R. Pimlott, *The Englishman's Holiday: A Social History* (Hassocks, 2nd edn, 1976), 21-64.

3    John F. Travis, 'The Rise of Holidaymaking on the Devon Coast, 1750-1900, with Particular Reference to Health and Entertainment' (unpublished Ph.D. thesis, University of Exeter, 1988), 348-51.

4    James Cartwright, *The Travels through England of Dr Richard Pococke during 1750, 1751 and Later Years* (1888), 102; Andrew Brice, *The Grand Gazeteer* (Exeter, 1759), 551, 1128.

5    Hugh Downman, *Infancy: Or the Management of Children* (Exeter, 6th edn, 1803), 117-8.

6    Devon Record Office, 564/M, John Swete, 'Picturesque Sketches of Devon', MS (1795) X, 159-60.

7    *Exeter Flying Post*, 2 August 1771.

8    Andrew Oliver, ed., *The Journal of Samuel Curwen, Loyalist* (Harvard, 1972), I, 204.

9    'Teignmouth', *The Royal Magazine*, VI (1762) 128. A reference by J.A. Bulley led me to this article: John A. Bulley, 'Teignmouth as a seaside resort before the coming of the railway', *Transactions of the Devonshire Association*, LXXXVIII (1958), 145-6.

10   *Exeter Flying Post*, 17 September 1789.

11   *Exeter Flying Post*, 6 September 1791.

12   *Exeter Flying Post*, 16 June 1791.

13   Travis, 'Rise of Holidaymaking', 391-406.

14   Thomas Rammell, *Report to the General Board of Health on ... the Sanitary Condition ... of Ilfracombe* (1850), 14, 16.

15   William White, *History, Gazetteer and Directory of Devon* (Shef-field, 1850), 446.

16   A.B. Granville, *The Spas of England and Principal Sea-Bathing Places* (1841), III, 483; Thomas Rammell, *Report to the General Board of Health on ... the Sanitary Condition ... of Torquay* (1850), 23.

17   *Western Times*, 27 August 1878; *Devon Weekly Times*, 30 August 1878.

18   Travis, 'Rise of Holidaymaking', 416-7.

19   Brice, *Grand Gazetteer*, 551-1128.

20   Downman, *Infancy*, 21, 116-8.

21   *Exeter Flying Post*, 2 August 1771, 6 September 1791.

22   John Ingenhousz, 'On the degree of salubrity of the common air at sea compared with that of the seashore and that of places far removed from the sea', *Royal Society: Philosophical Transactions*, LXX (1780), 375. A reference by John Whyman led me to this article: John Whyman, 'Aspects of Holidaymaking and Resort Development within the Isle of Thanet with Particular Reference to Margate, circa 1736 to circa 1840' (unpublished Ph.D. thesis, University of Kent, 1980), 136-7.

23   *Exeter Flying Post*, 17 June 1791.

24   *A Guide to All the Watering and Sea-Bathing Places* (1803 edn), 197.

25   John Evans, *The Juvenile Tourist* (1804), 31-2.

26   Edmund Butcher, *An Excursion from Sidmouth to Chester in the Summer of 1803* (1805), 450.

27   Octavian Blewitt, *Panorama of Torquay* (1832).

28   *Exeter Flying Post*, 4 June 1846.

29   Charles Hadfield, *Atmospheric Railways: A Victorian Venture in Silent Speed* (Gloucester, 1985), 143-76.

30   The population of Torquay (Tormohun parish) rose from only 838 in 1801 to 11,474 in 1851; Decennial Census.

31   Public Record Office, (PRO), MH13/184, 27 September 1849.

32   PRO, MH13/99, 19 October 1849.

33   *North Devon Journal*, 15 July 1875.

34   PRO, MH12/2145, 9 October 1878.

35 PRO, MH12/2145, Report of Lynton Medical Officer of Health for 1882; Lynton Town Hall, Lynton Local Board Minute Book 1881-5, 29 October 1884.

36 Travis, 'Rise of Holidaymaking', 395-435.

37 For an example of a guidebook describing the Devon coast in the stylised idiom of the picturesque school see, T.H. Williams, *Picturesque Excursions in Devonshire and Cornwall* (1804).

38 *Ilfracombe Chronicle*, 15 February 1873.

39 James J. Hissey, *On the Box Seat* (1886), 338-9.

40 L.J. Jennings, 'In the wilds of North Devon', *Murray's Magazine*, IV (1888), 81.

41 *Bideford Gazette*, 2 June 1863.

42 S. Baring Gould, *Devon* (7th edn, 1924).

43 For a full account of Philip Gosse's life see the account by his son: Edmund W. Gosse, *The Naturalist of the Seashore: The Life of Philip Henry Gosse* (1896).

44 'Off the rails', *Temple Bar*, VII (1836), 291.

45 Philip Gosse, *Land and Sea* (new edn, 1879), 251.

46 Edmund Gosse, *Father and Son: A Study of Two Temperaments* (1913), 130.

47 Gosse, *Father and Son*, 130.

48 Fanny Kingsley, ed., *Charles Kingsley: His Letters and Memories of His Life* (1883), 154.

49 Charles Kingsley, *The Water Babies* (new edn., 1878), 211-3.

50 W. Hyett, *A Description of the Watering Places on the South-East Coast of Devon* (Exeter, 1803), 83.

51 T.H. Cornish, *Sketch of the Rise and Progress of the Principal Towns of North Devon* (Bristol, 1828), 28.

52 *North Devon Journal*, 4 June, 18 June 1863.

53 *North Devon Journal*, 20 July 1876.

54 *A New Pictorial and Descriptive Guide to Torquay ... and Other South Devon Watering Places* (1898), x. A reference in F.B. May's thesis led me to this quotation: F.B. May, 'The Development of Ilfracombe as a Resort in the Nineteenth Century' (unpublished M.A. thesis, University of Exeter, 1978), 368-9.

55 Travis, 'Rise of Holidaymaking', 579-81.

56 *Torquay Times*, 7 September 1900.

57 *Torquay Directory*, 5 June 1890, 16 May 1894; *Western Times*, 2 August 1892; *Dawlish Gazette*, 9 June 1900.

58 John K. Walton, *The English Seaside Resort: A Social History 1750-1914* (Leicester, 1983), 71-2.

59 William Miller, *Our English Shores* (Edinburgh, 1888), 7.

60 *North Devon Journal*, 24 June 1849.

61 *North Devon Journal*, 25 August 1864.

62   *Ilfracombe Chronicle*, 22 August 1896.
63   Richard Warner, *A Walk through Some of the Western Counties of England* (Bath, 1800), 113-4, 120.
64   Williams, *Picturesque Excursions*, 43-4.
65   John Murray, *A Handbook for Travellers in Devon and Cornwall* (1851), 107.

# POLLUTION AND RESORT DEVELOPMENT: TEIGNMOUTH IN THE TWENTIETH CENTURY, AND ESPECIALLY IN RECENT YEARS

## John Channon

The prosperity of Devon since the eighteenth century has depended more and more on its holiday industry, the county's comparative advantage stemming from its natural beauty and the amenities of its coastlines. The contribution by John Travis in this volume on the nineteenth century has already reminded us of instances where man came to natural beauty and then despoiled it. Yet tourism has had an increasing significance for the local economy. Almost a third of all direct expenditure by holidaymakers remains as income to Devon residents while indirectly tourism creates business for service industries which provide employment.[1]

Various factors affect the number of tourists in Devon at any one time: repercussions (boom/recession) from the national economy; the vagaries of the weather; competition from overseas tourist centres; and, more recently, the development of inland 'activity' holidays. Tourism in Devon is also still highly seasonal with a concentration in July and August. This produces traffic congestion, uneven demands on local services, extra pressure on water supplies, high unemployment levels in winter while the benefits are poorly distributed throughout the county. Many of these factors are clearly beyond the control of the local holiday industry, yet other factors are of local origin and in recent years the central focus of attention has been the pollution of Devon's beaches and coastal waters.

For all the public outcry, examination of the resort's history reveals that, even though the nature of the problem may have changed, pollution itself is far from new. In the decades before 1914, for instance, several of the same issues affected the development of resorts. This paper will examine the theme of pollution and resort development by focusing on Teignmouth, on the south coast of Devon, briefly during the Edwardian period and then, in more detail, in recent years. It reveals that contemporary problems result both from the failure to address earlier problems and twentieth-century progress, and examines how effectively the major pollution issues affecting the resort are currently being tackled. Placing discussion within such a comparative perspective should help to shed light on the value of the historical approach.

## THE EDWARDIAN PERIOD

By the beginning of the twentieth century Teignmouth found itself in competition with other Devon resorts as well as wider afield: it had to ensure its attractions were publicised effectively, that the leisure amenities it provided were continually updated and steps were taken to eradicate features which might retard the resort's development such as health and sanitation. Whatever kind of resort Teignmouth was to become, the latter factors had to be addressed.

Before it could advance further as a resort, Teignmouth had to overcome the problem of health and sanitation. By far the most important sanitation issues concerned pollution in the River Teign and the need to improve Teignmouth's water supply.[2] Sewage disposal, from both Teignmouth and Newton Abbot, a town some six miles up-river, was the main problem as far as the Teign was concerned. Cases of typhoid were also reported after bathing in Teignmouth while the dangers of eating cockles and mussles caught from the river were regularly brought to light in the area's health reports.[3] Not only was this a health hazard for the local population - and a blow to the fishing community when threats were made to prohibit the sale of shellfish from Teignmouth - but also did little to enhance Teignmouth's reputation as a health resort and threatened to negate the positive work being undertaken to publicise the town.

The provision of an adequate and efficient supply of water all the year round took time. Twenty eight years elapsed between the initial discussion of the issue and its final accomplishment on 23 July 1908, a day described locally in quite extravagant terms as 'probably the most important date in the history of Teignmouth, an epoch in the annals of the town's history' since 'the one flaw that has hitherto stigmatised the fair name of Teignmouth' has been blotted out. Teignmouth and Paignton (further along the coast to the west) were now able to take a plentiful water supply from the uplands of Dartmoor via the Holne Moor (Venford) reservoir. A tourist resort obviously required an adequate supply of water during the summer season when, in addition to permanent residents who numbered 8,500, about 6-7,000 visitors descended on Teignmouth at any one time.

The long delay in realising the project was not due solely to prevarication. The conflict of interest over the various schemes proposed was a cause of delay. There were also physical difficulties that had to be overcome: to include Shaldon (a village on the opposite bank of the Teign from Teignmouth itself), the Teign had to be crossed in its

*Plate 1.* The Teign entering the English Channel. (Teignmouth to the left, and Shaldon on right bank), c.1970.

tidal range, while the district between Teignmouth and Paignton was obstructed by various clay mines, quarries and rivers. The work of laying pipes was often difficult because of rock and water in a number of sections. Apart from the approaches to the river Teign, the route ran along the existing service reservoirs at Hazeldown, Landscore and Shaldon.[4]

## THE LATE TWENTIETH CENTURY

Turning now to the late 1980s and very recent years, outdated and inadequate sewage disposal facilities can still be seen as a local failing, a recurrent problem caused by long-term deficiencies in the local infrastructure. Since such provision has been left for decades it has now become an exceedingly costly problem and, of course, modern communications allow for the more rapid spread of information as well as comment on any deficiencies through the media.

### Pollution and the Beach

Recent years have also produced cases of pollution of the Devon beaches by oil spillage. Teignmouth's beaches were only slightly affected by the *Rose Bay* incident in May 1990, when a supertaker collided with a Brixham-based trawler twelve miles south of Start Point and a large quantity of oil leaked from the tanker and was washed up on the beaches of Bigbury Bay. In August that same year three-quarters of the River Teign estuary below Shaldon Bridge was polluted with diesel oil from a cargo ship when it lost over fifty gallons while lightening its cargo of clay to cross the sand bar.[5] There were also reports of oil on Teignmouth and Dawlish beaches in mid-March 1991 explained by vessels clearing out their bilge tanks and spillage on the South Hams coast; in other cases oil was dragged up from the sea bed by rough water in high winds.[6]

An appreciation of the importance of beach and coastal pollution for the tourist industry has led to a number of campaigns in recent years to encourage resorts to improve the condition of beaches and inshore waters. A National Clean Beaches Award scheme was initiated in 1987 in conjunction with the 'Keep Britain Tidy/Better Beaches' campaign and Blue Flags were awarded to beaches which complied with the comparatively strict European Community (EC) standards while a national 'beaches week' was held in July.[7] Resort competition became the name of the game since favourable publicity clearly accrues from such awards (though, of course, the reverse is true for failure).[8] It was announced in early 1987 that more than one hundred resorts in the West Country, including Teignmouth and Dawlish, were to get a big clean up.[9] Shaldon beach failed a European test, the results of a two year survey, although Teignmouth and Ness Cove passed.[10] In 1987 Teignmouth was one of only two Devon resorts to win a 'clean beach' award from the Keep Britain Tidy Group though that same year Teignmouth beach was called a disgrace, dog fouling was reported and beaches failed the pollution test.[11]

Thirty one British beaches were contenders for the 1988 'Blue Flag' award (including Teignmouth) though some six Devon beaches (in Torbay) were amongst the seventeen successful ones. Teignmouth was not among these, and the judges did not even bother to visit it because of the poor state of its bathing water.[12] In addition, reports of sewage and other flushed waste matter floating in the water, paraphernalia of drug users found on the beach and such similar publicity did little to help the continued cultivation of the 'English Riviera' image. Some positive effects have emerged from this, however, such as the more stringent tests intitiated by the South West Water authorities on sea water quality while material improvements have included the construction of a screening plant at the Hope's Nose outfall near Torbay.[13] Both Teignmouth and Dawlish, however, did win other beach awards in 1988.[14]

In 1989, 74 beaches in England, Scotland and Wales won 'Tidy Beach' awards and Teignmouth was included. The judges looked at such facilities as litter-bin provision, accessibility of public toilets, and gave credit to local authorities which combined 'good management of beaches with innovative publicity campaigns' and which urged users to look after them.[15] Teignmouth won a 'Blue Flag' in 1990, though Holcombe beach nearer to the sewage outfall failed. Beach boards informing swimmers of water quality must be displayed at Blue Flag beaches and updated throughout the season.[16] On 11 July 1991 Teignmouth holidaymakers and schoolchildren joined in celebrations having watched the Blue Flag being unfurled on the seafront.[17] This was the second year that Teignmouth had won the award and at the time it was the only beach in Teignbridge to do so. Certainly the dog bans during the summer season ensure the eligibility of the beaches for blue flags in the future, though it was feared that the state of the River Teign and of the Royal Hotel - the latter's run down condition a prime example of visual pollution - might place it in jeopardy.[18]

Doubt, however, has been cast on just how 'clean' a 'Blue Flag' beach is since only bacterial tests are included. Viral tests, argue some experts, are just as necessary.[19] Pollution was still present on Teignmouth beach and a hazard to swimmers in the summer of 1990 and a Green Party survey also argued that beaches with Blue Flag status (including Teignmouth, Dawlish and Exmouth) were badly pol-luted.[20] The number of beaches in the South West which failed cleanli-ness tests doubled in 1991, four in South Devon including Shaldon and

Ness Cove. Beaches are tested forty times a year and Shaldon failed five times and the Ness four. In past years these two also failed though the Ness only for the second time.[21] The 1992 Blue Flag standard was to be raised with a call for South West Water to bring forward its programme for treating sewage.

Already in early December 1991 the local press was reporting that the new Teignbridge holiday guide would be dropping a photograph of the Blue Flag at Dawlish Warren (the resort lost the award due to a freak tide of sewage striking the sand on the very day the judges arrived) and, although it was too late to withdraw a similar photograph at Teignmouth, work on the River Teign sewerage scheme would lead Teignbridge to refrain from entering the resort in 1992. The Teignbridge Recreation and Tourism Officer admitted that possessing the Blue Flag provided useful marketing and was of great benefit to resorts; holidaymakers did take notice of it.[22]

This was followed in early February 1992 with a report from the National Rivers Authority, which samples water at beaches around the country, that more beaches in the South West failed during 1991 than in the previous two years, including Ness Cove at Shaldon. The reason was untreated sewage discharged close inshore and washed back onto sands at popular resorts. It was accepted that water quality in the region has deteriorated but the main cause of the failures was believed to be the difference in sunshine levels during the summer with the sun shining significantly longer in 1989 and 1990 than 1991, killing off more sewage-derived bacteria in the water.[23] Both Shaldon and the Ness Cove failed the tests (while Dawlish was relegated from four to three stars, the same as Teignmouth and Holcombe, because of sewage problems) to determine which resorts would be included in the 1992 edition of the *Good Beach Guide*: and if waters failed, it warned that swimming and other immersion sports could cause illness. According to the Teignbridge Recreation and Tourism officer, the massive scheme to clean up pollution in the Teign estuary should improve the water quality of all beaches in the area by 1994.[24] The Marine Conservation Society examines over 450 beaches in Britain of which 155 are in a recommended section.[25] Those recommended have consistently good water quality and few problems from sewage debris. EC bathing water directives have to be achieved by 1995 and water companies have had to respond accordingly with new investment programmes. The 'best of the best beaches' have passed the very strict EC bathing waters directive guideline coliform standards as opposed to the much less strict minimum standards.

## Pollution of the river Teign

Some of Teignmouth's problems are common to resorts throughout the region. At the height of the summer some half of all sewage created by the South West's population (half a million visitors at any one time) pours untreated into the sea while every day the sewage of three quarters of the resident population follows the same route. Some is untreated crude sewage, some is effluent from treatment works and some is macerated screened sewage. The South West Water (SWA) plans treatment works in Torbay so that by the year 2000 no untreated sewage is pumped into the bay. The use of the sea as a method of sewage treatment is one reason why there can never be a guarantee that bathers off the beaches can be sure that they are entirely free of bacteria and viruses.[26]

Other problems are more specific to the locality. Over the years the Teign has become increasingly polluted with sewage and other debris; reports of pollution in the Teign were present in the early 1960s.[27] More recently the Teign has been classed as a health hazard for bathers and water sports enthusiasts and the tourist industry suffered from the unedifying sight of raw sewage washed up on beaches at Teignmouth and Shaldon.[28] Sewage disposal problems and pollution on the beach and in the estuary have had detrimental effects on the holiday trade.[29]

Concern was expressed in the local press in October 1989 about reports of serious pollution of the Teign,[30] while the Chief Environmental Officer at Teignbridge issued a warning to avoid using the river for sailing or swimming, since the raw sewage being pumped directly into the Teign at various points along the river, could produce viral infections. The problem is a combination of public and private sewage: some sewage comes from Buckland treatment plant at Newton Abbot, some from the Coombe Cellars area above Teignmouth (especially the public house where a new plant was being installed). The Teign is shallow and does not benefit from dilution as do larger rivers. Usually sewage is released on the ebb tide but occasionally the sea will flow upstream at lunchtime. The old outfall pipe in Shaldon became blocked with sand in 1987 and had to be replaced by a new pipe (when the old one was unblocked sewage would seep through, become trapped on rocks and brought back into the river by the incoming tide).[31] A warning came at the Institution of Environmental Health Oficers annual conference in 1990 that swimming in many of Britain's 'clean' Blue Flag beaches (seven of which were in Devon that year and included Teignmouth and Dawlish Warren) could bring bathers into contact with viruses linked to meningitis, hepatitis and paralytic disorders, even though the risk was small. 140 beaches in the country were the subject of European dirty water prosecutions that

RIVER TEIGN SHOWING (SOUTH WEST WATER) TEIGN ESTUARY COASTAL SEWAGE TREATMENT IMPROVEMENT SCHEME, PROPOSED MARINA DEVELOPMENT ON THE SALTY AND TEIGNMOUTH AND SHALDON BEACHES

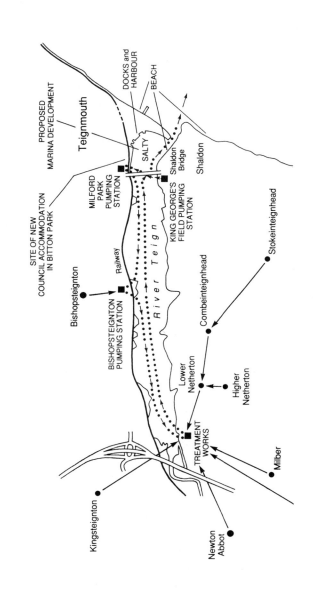

year. All of this was sufficient to heighten awareness of the worsening problem and increase concern to remedy it.[32]

As far as health in the Teign is concerned, food poisoning especially from shellfish is a major concern which in turn affects marine life.[33] This, as we saw above, was also an issue in the Edwardian period. A new attempt was embarked upon in October 1986 to improve the quality of oysters in the Teign, some five years before the council ordered that all shellfish grown in the Teign could not be sold until they had been relaid in purified water for four weeks since the shallow water of the Teign was rarely washed out, thus allowing the growth of viruses.[34] Oysters were linked to food poisoning in late 1987 and a warning issued in early 1988 while in response Teign mussels were declared safe a week later.[35] There was also conflict in 1988 over the possibility that some of the thousands of tons of ammonium nitrate (fertiliser) imported into Teignmouth in early May that year, and stored in bags on the quays, had been washed into the Teign posing a threat to fish and shellfish, though this was officially denied. Even if such reports proved baseless, they served as further evidence of the heightened perceptions locally of the dangers of pollution.[36] New rules were introduced for the oyster industry after an increased incidence of poisoning from the Teign[37] while some half a million mussels, harvested from the Teign each year, are cleaned out by the tidal waters of the estuary before going through more thorough purification ashore.[38] In the light of history there is a certain irony in the words of a Teign oyster grower in October 1991 who, concerned to emphasise that oysters were 'perfectly safe to eat' declared that the situation is 'exactly the same today as it was for the last hundred years'.[39]

The scheme to rectify these problems envisages that raw sewage from Teignmouth and Shaldon currently flushed straight into the river will be pumped upstream and treated at Buckland (see Map). It will then pass back down again, along with sewage from Newton Abbot and the Barton new town (Torquay) and out to sea off Shaldon via a two kilometre outfall pipe. Although some public concern has been expressed that the length of such a pipe is insufficient, and that pollution may still be washed back onto the beaches, specialists at South West Water (SWW) have explained that a longer pipe would have no advantages: there will be no solids in the sewage and the coldess of the water, the salt and the sunlight will quickly kill off most bacteria. Disinfectant would have the same effect but this is not allowed because the chemicals would be harmful to marine life.[40]

The sewage problem has been allowed to build up over many decades and even though a massive residential building programme has been undertaken in the area over recent decades, effective action has not been forthcoming. The antiquated sewerage system is to be modernised for the first time in more than a century - a spin off from the SWW scheme to clean up the polluted River Teign - the old pipes having barely been touched since laid by the Victorians. A massive new sewer main, linking up with all the old sewers running down the valleys, is to be laid right through the heart of Teignmouth.[41] At the moment all pipes flow to the Gales Hill storage tank on the harbour but the new pipe will divert all the sewage and storm water to a brand new pumping station at Milford Park.[42] Then all sewage from Milford and from storage tanks at Shaldon and Bishopsteignton will be pumped up to Buckland where it will be fully treated and then sent back downstream via another pipe and deposited at sea through an outfall pipe at the Ness.[43] Although the storage tank at Shaldon will be completely buried, a further problem has arisen because it has only recently been revealed that the overflow screen chamber to relieve pressure in the tank has to go on top, above ground. Assurances have been given, however, that the chamber will not be an eyesore because it will be grassed over.[44] Work on a construction platform at the Ness began in early 1992 and sections of pipe for the outfall are to be welded together on the platform and then pulled into the sea.[45] Pipes will be taken out to sea by barge lowered into a trench on the sea bed and connected by divers.[46]

South West Water is spending some £30 million (one of the biggest capital investments ever in the area) on 'Operation Clean Sweep', cleaning up the river Teign, and ridding it of sewage, although the cost will mean increased water bills for local residents. The project is due to be completed by summer 1993.[47] The new public outfalls, controlled by the SWW, should mean greatly improved discharge into the Teign (meeting European Community standards). Private sewage outfalls may continue to release raw effluent and even though the Teign as a large body of water can absorb a certain amount of outfall, watchdogs such as the National Rivers Authority (NRA) are to take action to encourage owners of the many small private discharges to improve matters. Another option would be to link into the public sewer though this would be costly.[48]

The Ness outfall will be scrapped and sewage treated at a new works before ending up in the sea through a long outfall - pumped up to Buckland and then back down the Teign to the sea.[49] New pumping stations and storage tanks are being built at Shaldon, Teignmouth (Milford Park) and Bishopsteignton, while the capacity of the Treatment works at Buckland, Newton Abbot, is being doubled.[50] To cope with the deep and spongy mud of the Teign two American amphibious diggers are being used - the latest state of the art technology - the only ones of their kind being employed in Europe, while the ten miles of plastic pipeline shipped into Teignmouth docks from Denmark and Portugal has made it the largest collection of pipes ever assembled in the U.K., and provided good publicity in itself for the area.[51] At the same time the project has already attracted interest from overseas countries interested in environmental issues, providing a possibility for selling such expertise abroad in the future.[52] Barges have also made a return to the Teign since they are the best way of transporting materials and equipment around the various riverside sites.[53]

In addition to the SWW plan to remove sewage, another plan to clean up the Teign catchment area (from Dartmoor to the sea) was announced by the NRA, its aims being to ensure water quality objectives are met, maintain fish stocks and enforce pollution legislation (affecting the various metals mined in the river as well as the quarrying of stone and ball clay). The area is also an important source of drinking water with several reservoirs and good runs of salmon and sea trout.[54]

The South West NRA in September 1991 opened a new £4 million pound laboratory next to the NRA headquarters at Sowton, near Exeter, with the aim of providing a chemical and microbiological analysis of water from rivers, streams and beaches around the South West. (In all, it will deal with bacteriological tests on bathing water, chemical analysis of river water and effluent discharge, organic tests for pesticides and analysis of metals.) The section carrying out tests for water born viruses is the only one of its kind in the country. Samples will be taken from beaches to see if they meet the EC standards for water quality. Twenty samples are taken from each bathing beach between May and September to check for traces of pollution and every beach will be checked twice during that period for signs of salmonella and various viruses.[55]

The question of pollution has also centred around two other issues: plans to build a marina and the proposed changes to Shaldon bridge. A marina-housing scheme on the Salty island in the estuary (or an alternative location near the sea wall in Teignmouth) in mid-1988 (see Map) produced considerable local opposition. Not only was it seen as an eye-sore (visual pollution) spoiling the natural scenery of the Teign estuary but the environmental hazards to marine life were stressed as well as the effect of the shifting sand bar on flooding and water levels.[56] Opposition was on a number of fronts but especially on environmental grounds and the impact on the local economy. It was considered that the hydrological effect on the estuary would be disastrous; as would be the traffic escalation; the loss of boatmen's moorings together with most of Shaldon's beach; while the livelihood of fishermen and musselmen would also be threatened. The character of Shaldon would be destroyed and the beauty of the estuary spoiled.[57]

As had been the case with the construction of council homes on Bitton Park the question of ownership and user rights was raised, including the questioning of who owns the Salty and whether the council has the right to sell it. This also saw the re-emergence of an issue first mooted in the 1960s.[58]

In April 1991 a £7 million plan was unveiled to bring the sixty-year-old Shaldon bridge up to EC standards whereby its current weight limit of 38 tons will need to be increased to 40 tons by 1999.[59] A structural survey revealed weaknesses and so weight restrictions have been imposed.[60] It was proposed to demolish the bridge and build a new, taller bridge. On the plus side, this would enable an increased number of larger boats to navigate the higher reaches of the Teign, bringing benefits to tourism (though it has been questioned whether a higher bridge would necessarily allow more tourist development on the river). It could mean crossing the river at a different angle near Inverteign and thus provide an opportunity of creating a new access road to the docks.[61] On the debit side, it would spoil the views, from the estuary towards Teignmouth and from Teignmouth towards Dartmoor. The beauty of the estuary is that it is still relatively unspoilt (and a reason for tourist visits ) and a haven for a great variety of birds (providing an essential feeding and roosting area). Spoiling the environment in this way might well have adverse effects on the holiday industry.[62]

**Pollution more generally**
Both the marina and Shaldon Bridge controversies provide good examples of concern about a wide range of pollution issues. In general the second half of the twentieth century has seen an upsurge of noise, visual and cultural pollution. The internal combustion engine and new forms of leisure activity have led to health risks (from exhaust fumes), more cars and coaches and hence more land being given over to car parks, with street bans on cars and recently the pedestrianisation of a part of the town centre. This is quite apart from the increased spending on all this as well as on roads and other transport infrastructure.

Visual pollution has also been of growing concern locally. Environmentalists failed to convince councillors to halt the plan to build thirty one sheltered homes on a site in Bitton Park, an area of outstanding beauty, notwithstanding the fact that this was public land.[63] Bitton Park had been left to the people of Teignmouth almost ninety years ago and handed over to Teignbridge on local government reorganisation in 1974. The opponents to the housing scheme claimed that it would encroach on a large part of the park, with the loss of trees and estuary views.[64] The scheme, it was claimed, would ruin one of the last few remaining green areas in the town, and it, together with the growth of the local docks towards Shaldon Bridge, would spoil the shore of the Teign estuary (and was anyway contrary to the Local Plan). This controversy was and is indicative of the ever-present conflict of interest over the use of scarce resources. The council meanwhile argues that it is desperately short of council-run sheltered accommodation, in an area with an ever-ageing population.[65]

The fear of encroaching visual pollution is ever present as witnessed by two reports early in 1992. Plans for a childrens' adventure holiday centre at Old Teignharvey, Combeinteignhead, raised fears that this would prove an eyesore for those looking across the Teign from Bishopsteignton - though this was given the go-ahead by Teignbridge Planning Committee:[66] while local opposition has emerged to plans to build a house and garage on the headlands (above the 'Parson and Clerk') at Holcombe between Teignmouth and Dawlish. The Teignmouth and Shaldon Environment Society has pointed out that the headland is in a Coastal Preservation Area, that it is an area of Great Landscape Value and the cliff is a site of Special Scientific Interest. Fears were raised for the landscape and the wildlife habitat.[67] Also of concern has been the construction of a new Dock warehouse on the Eastern Quay in Teignmouth (a visual blight to the residents opposite;[68] and the decrepit state into which the London Hotel - once the most prestigious hotel in the resort (and host to the Beatles amongst others in the mid-1960s) - had been allowed to sink. Environmentalists also opposed a plan to build forty four houses on the former holiday camp site (Teignhaven Holiday village) on the river edge near Shaldon Bridge, though this was presumably quashed with the Teign Estuary Scheme.[69]

It becomes clear that the various forms and types of pollution are interlinked. Teignmouth like many resorts has had to cope with the associated effects of the internal combustion engine - exhaust fumes and carbon dioxide poisoning together with sulphur dioxide emissions - and more recently, acid rain. At the beginning of the century complaints were heard about visual, noise and health pollution from railways in a similar way. Dumping of toxic waste has affected both the Teign and the sea and beach area as has sewage and water supply. Oil spillages have had a similar impact. In addition the beach area is affected by dog fouling, dropped litter and drug associated paraphernailia. Oiled birds, litter, household refuse and toxic industrial waste were the most serious problems reported in the Norwich Union Coastwatch survey published in February 1991.[70]

Environmentally, one can see the effects on the marine environment of town growth: polluted air (smoke and exhaust fumes; industrial and recreational developments, such as the marina), and new residential developments such as flats. These have been also a cause of visual pollution, destroying parts of the coastal scenery. Noise pollution has become an increasing problem over the century, from brass bands to transistor radios and more recently to modern portables such as ghetto blasters. To some residents, traffic, music and amusement arcades also produce cultural pollution, though this is clearly a subjective matter, and a product of observer bias. It does, however, define the 'tone' of a resort where age divisions have become increasingly important.[71] As we progress to ever newer forms of technology so these problems become more acute. The recent phenomenon of jet skis, for instance, produces more noise and petrol in the water, and can pose dangers to bathers, particularly small children. Safety relies on the users acting responsibly. As society progresses technologically so such problems become all the more acute and such developments require even greater consultation between resource-users.

As to the future, the Teignbridge *Local Plan*, drafted in July 1991, was the blueprint for a 'Greener Teignmouth' against the background of increasing public concern for the environment. It argued that Teignmouth was struggling to maintain its role as a resort, suffering from a lack of investment and such problems as summer car parking. Many recommendations related to roads, pedestrianised areas and sanitation, though, importantly, it noted that no new tourist facilities were to come from the council.[72] Protection of the environment was to be given top priority, a major concern being the unprecedented growth in building activity over the past two decades which, while providing health and social benefits, had raised public concern about the environmental implications of growth on such a scale. The Plan also supported the protection of Coastal Protection Areas and Areas of Great Landscape value, with other policies that were to be aimed at protecting areas of countryside and the urban fringe. Perhaps one of the most significant provisions was that no new 'greenfield' sites were to be allocated for housing in the Plan.[73] As to tourism, the Council's aim was to 'maintain the character of the area as one suitable for family holiday enjoyment', and seeking to discourage any further loss of holiday accommodation in the resort.[74] Measures were envisaged to protect areas of nature conservation and landscape value; to use traditional materials; and to control advertisements and appropriate designs for shopfronts - all important factors for ensuring that any further visual pollution would be kept to a minimum.

## Conclusion

It must be concluded, therefore, that Teignmouth like most resorts has to work to reduce the negative factors (especially pollution of the beach bathing areas and the river Teign) which make it less competitive. This is particularly important with the fall-off in tourism that has occurred since 1991, while pollution, and the bad publicity associated with it, are now high on the agenda as the resort approaches the twenty first century, just as it was at the turn of the last century. The South West Water 'Clean up' scheme, together with other measures such as those of the National Rivers Authority, should produce cleaner beaches and purer rivers, though external pressures (such as improvements necessary to meet more stringent European Community standards) have given increased urgency to realising such change. Perhaps the achievement of the provision of an adequate and efficient supply of clean water all the year round in 1908 will be paralleled in the mid-1990s by the successful completion of the Teign Estuary Regional Sewage Disposal Scheme. In Edwardian times there were complaints that it had taken twenty eight years from the initial discussion to complete the water supply project. It will have taken much longer to realise investment in a new sewage scheme (and the same factors, procrastination and rising costs, have been largely to blame). But if, as anticipated, this will eradicate, or at least reduce considerably, pollution both in the Teign and on the local beaches, the benefit to the town's resident population and to the resort's future will be as great, if not greater, than that of almost a century ago.

## NOTES

1   *Devon in Figures* (Devon County Council, 1985), chap.10 and p.46.
2   *Teignmouth Post* (hereafter *TP*), 6 May and 8 July 1904; 5 June and 24 July 1908
3   *TP*, 7 October 1904; 7 July 1911; 8 March 1912; 31 July 1913. For the dangers to bathing from sewage see, *TP*, 18 March 1904; 30 March 1906.
4   For details see John Channon, 'Teignmouth as a resort before the First World War', *Maritime South West*, 4 (1988), 28-36.
5   *Express & Echo* (hereafter *EE*), 8 August 1990. The tanker was Liberian registered.
6   *TP*, 15 March 1991.
7   The NRA carries out the water quality tests used in the designation of Euro and Blue Flag beaches. For details of how the tests are carried out see *Herald & Express* (hereafter *HE*), 17 September 1991.
8   For a brief summary of how these relate to other Devon resorts see John Channon, 'Seaside tourism: The pollution problem', in Michael Duffy, et al, eds., *The New Maritime History of Devon*, Vol. II, Conway Maritime Press, London, 1994), chap. 26.
9   *TP*, 6 February 1987.
10  *TP*, 8 May 1987.
11  *TP*, 31 July and 11 September 1987. For 'disgrace' see *HE*, 19 March 1987; dog fouling, *TP*, 17 July 1987; and pollution test, *HE*, 12 November 1987.
12  The appropriateness of some of the awards was questioned anyway. The residents at Meadfoot, for instance, claimed that the beach was full of broken glass, rubble and scrap metal.
13  Channon, 'Seaside tourism'.
14  *HE*, 12 August 1988.
15  *The Guardian*, 31 July 1989.
16  *HE*, 10 July 1990.
17  The award lasted until the end of May 1992, *HE*, 3 December 1991.
18  *HE*, 12 July 1991.
19  For discussion see, *Teignmouth News*, 15 August 1991.
20  *HE*, 10 July 1990, 17 August 1990.
21  *HE*, 13 November 1991.
22  The sewage striking Dawlish Warren possibly came from as far away as Weymouth, a good illustration of the difficulties affecting resorts. It is expected that the Teign sewerage works will be completed within a year. *HE*, 3 December 1991. Only seventeen British beaches were awarded the Blue Flag in 1992, four of which

(Seaton, Meadfoot, Oddicombe and Woolacombe) were in Devon. *The Guardian*, 5 July 1992.

23   *HE*, 1 February 1992.

24   *HE*, 3 March 1992. The *Good Beaches Guide* is produced by Heinz in conjunction with the Marine Conservation Society.

25   *Western Morning News* (hereafter *WMN*), 5 March 1992.

26   *HE*, 17 September 1991.

27   See, for example, *TP*, 25 August 1961.

28   *HE*, 16 June 1987.

29   *TP*, 5 December 1986. For pollution in the Teign see, *TP*, 17 October 1986; sewage and pollution in the Teign, *HE*, 6 May 1988; sewage disposal, *HE*, 17 June and 7 December 1987; *EE*, June 1987; sewage outfall in the Teign, *HE*, 25 January 1988. Local swimmers and holidaymakers complained, claiming that they would avoid the river in future.

30   *HE*, 25 October 1989.

31   *HE*, 16 June 1987.

32   *HE*, 18 September 1990. The warning came from a water expert from Surrey University.

33   These have also been affected by the works at Teignhaven.

34   *TP*, 17 October 1986.

35   *TP*, 13 November 1987; *HE*, 11 January 1988; *HE*, 18 January 1988.

36   *HE*, 30 May 1988. The denial came from the Teignmouth Quay Company.

37   *HE*, 6 July 1989.

38   *HE*, 25 October 1989.

39   *TP*, 25 October 1991.

40   *HE*, 27 September 1991. Attempts are still in progress to discover new chemicals that would not have such deleterious effects.

41   The sewer will be up to seven feet in diameter and almost a mile long and between 25 and 80 feet below ground.

42   Some effluent from the lower part of town will continue to run to Gales Hill where it will be transferred via another pipe to Milford Park.

43   *HE*, 20 November 1991. For details of the Bitton House/Rugby Ground sewer see, *WMN*, 6 March 1992.

44   *HE*, 29 February 1992. The chamber is to be two metres high, 20 metres wide and 10 metres long. It also means the loss of a space intended for a children's play area - due to be built on the site when the tank is completed - to replace one which had been displaced by the tank.

45   *HE*, 28 December 1991.

46   *HE*, 9 March 1992. Pipes will be laid at Flow Pit to Bishops Rock at Bishopsteignton alongside Shaldon Bridge and from the Ness

to Shaldon Bridge, *HE*, 27 December 1991.
47   *WMN*, 20 April and 7 February 1990.
48   *WMN*, 30 July 1991; *HE*, 27 September 1991.
49   *HE*, 10 July 1990. Holcombe was not on the 'clean up' list.
50   The trenches are located off the King George V playing field and the former neighbouring Teignhaven holiday camp serves as a base for the contractors.
51   *HE*, 16 September 1991. The pipes are to be shipped into the docks on cargo boats and then transferred to waiting lorries.
52   *HE*, 25 November 1991. The pipes will be buried out of sight in the mud and sand of the estuary.
53   *HE*, 27 September 1991.
54   *HE*, 14 November 1991. The impact of drainage from the A30 trunk road is also being assessed.
55   *HE*, 17 September 1991.
56   For marina proposals, see the local press for the spring and summer of 1988.
57   For opposition see, *EE*, 20 August 1988 and *HE*, 6 September 1988. Both the Shaldon Parish Council and the Teignmouth and Shaldon Environment Society came out against the scheme (see for example their leaflet 'Save the Salty!').
58   For further discussion of the marina proposals and controversy see, local papers, *TP*, and *TN*, summer 1988.
59   *HE*, 31 January 1992.
60   *HE*, 31 January and 5 February 1992.
61   *HE*, 3 February and 5 March 1992; *WMN*, 1 and 4 February 1992.
62   *HE*, 1 February and 25 March 1992.
63   For a favourable report see, *HE*, 14 December 1991.
64   *HE*, 28 March 1990.
65   See, *HE*, 24 April 1990; *WMN*, 24 and 30 April 1990, for Bitton Park flats; for *Teignbridge Local Plan*, see, *HE*, 4 July 1991 and *WMN*, 5 July 1991.
66   *HE*, 5 February 1992.
67   *HE*, 13 February 1992.
68   *WMN*, 5 November 1991.
69   *HE*, 17 September 1990.
70   The erosion of buildings is just one effect of acid rain about which concern has been expressed.
71   For an interesting comment see, D. St John Thomas, 'Tourism must not be taken for granted', in *Western Morning News: Selections from The Western Morning News, 1984-85* (Dartmouth, 1985). Proposals for a marina and other associated developments would clearly act to create a different 'tone' for Teignmouth as a resort, moving away from its 'tripper' image established earlier this century in an attempt to appeal to a more up-market clientele.

Whether Teignmouth could compete with other Devon resorts which have already developed marinas (such as Dartmouth and Torquay) remains to be seen.

72 For further details, see the *Teignbridge Local Plan*.

73 See, *The District*, issue 5 February 1992, 2.

74 See, for instance, the debate concerning the conversion of hotels and boarding houses into residential homes, luxury apartments, etc. in John Channon, 'Seaside tourism in twentieth-century Devon' (unpublished paper, 1990), 17-18.

# THE SEAMAN'S CONCEPTION OF HIMSELF

## Kim Montin

I wish to discuss a number of matters pertaining to the modern seaman. 'The Modern Seaman' is Åbo Akademi Maritime Museum's contribution to 'The Maritime Man' project, currently being undertaken at the Department of Ethnology at Åbo Akademi University.[1] The project falls within the discipline of maritime ethnology, and the observations below indicate some of the approaches contained in my own contribution to the project.

My prime concern is with the seaman's conception of himself, his work, and his environment. The intention is to achieve an 'approach from within' which in this particular case is called an 'emic approach'. I wish to develop an understanding of the attitudes, motives, interests, responses, conflicts and personal development of seamen.[2] The intention is to pursue a comparative approach, to consider seamen from the sailing ship era as well as the more modern, powered period. Three time periods at thirty-year intervals have been chosen, each interval corresponding to a certain type of vessel, that is, sailing vessels in the 1920s and 1930s, steamers in the 1950s and 1960s, and motor-driven vessels in the 1990s.

The study will be limited to Finnish seamen on Finnish cargo ships, mainly on routes in the North Sea and the Baltic. This particular qualification had to be relatively strictly adhered to as the living conditions of individual seamen vary a great deal dependent upon a vessel's type and size and the nature of the shipping routes it sailed. It would, in other words, not be very expedient to speak at one and the same time of a seaman on a passenger line between Sweden and Finland, or on a tanker which plies between Finland and the Middle East, as well as a sailor on a little cargo ship in coastal traffic.

A model for the study is to be found in Knut Weibust's Stockholm doctoral dissertation of 1969.[3] In this ethnological investigation Weibust discussed the miniature society made up by the crew of a sailing vessel. Central terms used by Weibust include 'role', 'norm', and 'status'. I have availed myself of Weibust's frame of reference, but while Weibust's study treats the sailing vessel alone, my aim is also to include steamers and modern cargo ships. There are also notable differences in the source material I will make use of. Whereas Weibust

*Plate 1. M.V. Winden.*

used literary sources alone, my intention is also to make use of questionnaires and interview material as well as 'participatory observation'. To this end I have already carried out some ethnological fieldwork on board the Finnish motor-driven vessel, *Winden*, a 3826 gross tons, lift on-lift off vessel built in 1986, on a two-week voyage. The voyage began in Kotka on the Gulf of Finland, proceeded through the Danish Sound to Grangemouth in Scotland, from there along the east coast of England to Purfleet on the Thames, from there across the North Sea to Rotterdam and, finally, through the Kiel Canal to Helsinki.

It might be added, that the experience of such a voyage suggests that every maritime historian and maritime ethnologist would greatly benefit from a similar experience. No matter how keen an amateur sailor an academic investigator might be, there is nothing quite like experiencing a modern ship at work in order to really understand what goes on. One learns a tremendous amount that would be impossible to acquire from reading about the experience alone; while at the same time many prejudices and stereotypes about sailors are inevitably shed. This was, for me, a strong personal experience, becoming aware of the crew and the vessel in a variety of conditions, as in darkness, stormy weather and dense fog.

The documentary material collected was compiled on board the *Winden* during the voyage, and in dock. It consisted of some 227 black and white photographs, 120 slides, approximately three hours of video recordings, about seven hours of recorded interviews and 52 pages of notes, as well as photostat copies of various of the vessel's documents. The material now forms part of the Åbo Akademi's Maritime Museum's primary historical holdings.

## The Seaman and His Image of His Vocation

Is the occupation of a seaman today still 'a hard and rough-mannered trade', a 'rootless existence with liquor interchanged with port brothels as sole forms of recreation', as suggested by a fiction writer in 1977?[4] Is the seaman's choice of profession today like that of the Ålander, Erik, who at the beginning of this century had to choose between two alternatives, to become a seaman or a farm-hand? His reasoning went like this: 'Seasickness doesn't last but the smell of dung remains for life'.[5]

To which social class does the modern seaman actually belong? Can he still be regarded as poor? This was still very much the case at the end of the Second World War, as a Finnish trade union official expressed it in reporting a meeting with the Finnish Prime Minister in 1946:

I tried to describe [to the Prime Minister] how Finnish sailors roam the streets of foreign harbours in threads because they can't afford better clothing. Nobody can support a family on 4500 FIM a month and that is why the wife and children of a sailor are inadequately clothed and often lack sufficient food. Hungry are the children of sailors when they lay their heads to sleep though their fathers toil for their sustenance.[6]

No, things are, as we all know, different today. Today seamen, certainly contemporary seamen in Finland at least, cannot be seen as poor. Apart from significant improvements in real wages there are many forms of recreation for seamen other than liquor and brothels. Besides the seamen's missions and churches (which, of course, existed before) a variety of other organisations have arisen to make the lot of sailors more acceptable, such as the Seaman's Service Agency which exists to offer different forms of recreational activities for seafarers, including sports, photographic competitions and various cultural events. Many modern vessels now have exercise facilities on board, as well as saunas, television and video and stereo equipment.

It is noteworthy how the modern Finnish sailor, despite these improved facilities on board vessels and ashore, suffers as much from feelings of stress and psychosomatic disorders of various kinds as he used to do from problems involving the abuse of alcohol. It is striking that in 1991 20 per cent of all sailors' invalidity pensions in Finland were granted on grounds of psychological illness.[7]

**The Characteristics of Seamen (or What Makes a Seaman a Seaman?)**

What, one wonders, constitutes the 'representative seaman'? What is it that makes him different from other people? Is there a grain of truth in the stereotypes we have of a sailor? Is it the way a seaman dresses, or his behaviour, or perhaps the other 'signals' he gives off which reveal him for what he is? Is it still, as was the case 150 years ago, as described by Richard Henry Dana Jr., in his *Two Years Before the Mast*:

A sailor has a peculiar cut to his clothes, and a way of wearing them which a green hand can never get .... Beside the points in my dress which were out of the way, doubtless my complexion and hands were enough to distinguish me from the regular *salt*, who, with a sunburnt cheek, wide step, and rolling gait, swings his bronzed and toughened hands athwart-ships, half-open, as though just ready to grasp a rope.[8]

Knut Weibust named the following 'attributes which provide cues or pointers facilitating the perception of the sailing ship sailor':

- his broad-bladed sheath knife
- his tattoo
- the sailor's characteristic rolling gait
- sailors' role-names or nick-names (like Jack Tar, Old Salt, Sea Dog, and so on)

Can these 'signals' still be detected among modern sailors? On board the motor vessel, *Winden*, I discovered that several of its crew members had such a knife in their possession, even those who could not be said to really need one in the course of their work, as did the boatswain. Moreover, a number of crew members did indeed have tattoos. (It may be of interest to note that the men were somewhat embarrassed when I asked questions about their tattoos in the *sauna* of the ship. The point was that they would not normally show their tattoos to a non-sailor for risk of it being assumed they had at some time been in prison). I could not, however, detect any sort of 'rolling gait' in the way they walked, nor could I find evidence for the use of nick-names of the sort suggested by Weibust. This could, though be explained by the limited size of my source material.

At this point it might be appropriate to consider what 'traditional things' one finds on board a modern seafaring vessel. If we disregard facts such as that commercial shipping has an extremely long history in the North Sea and Baltic Sea regions, that machine and propeller-driven vessels have also been around for a long time, that the handling of cargo is traditionally done with harbour cranes (the lift on-lift off system), and so on, we can quickly advance to the numerous details on board the ship. It is in a way surprising to find on a modern vessel like *Winden*, a pilot ladder which is an exact duplicate of those found in the sailing ship era, made of ropes and equipped with wooden steps. Mooring ropes and cable wire are spliced here as they were in the nineteenth century; shrouds and rigging screws which support the little mast at the bow are not exactly new inventions either. Regarding the navigational equipment, besides the electronic equipment which is of course in daily use, a magnetic compass, a mechanical chronometer and a sextant can still be found on board. Further, regardless of the modern communications apparatus, traditional signal flags can be found in a cupboard on the bridge.

Turning now to the matter of 'roles'. In order to fulfil his role as a sailor those who choose this profession must live up to the expectations that are placed upon them by the miniature society on board a ship. These expectations are expressed in the form of 'norms'. It is noteworthy that it is precisely the small size of the ship's community, 'the closed room', which strengthens role and norm consciousness. Weibust defines the term 'norm' as 'a socially sanctioned behavioural rule'. According to these norms sailors must, for example, keep their vessel 'ship shape' and obey their superiors. There are also norms governing working pace on board and others dealing with 'privacy'. It is, thus, of interest to analyse the expectations that are placed on a sailor and the expectations that they place on their environment, what they are expected to do and where they are expected to live and move about on board.

The roles played on board ship can be examined from several different perspectives, for instance, from the points of view of the different work categories. What is the role of a simple sailor, and what is the first officer's role? The role of gender on board can also be discussed: what is the role of women, for example? At this point, I would like to connect this discussion of roles and norms together with the different parts of the ship. The seaman's 'conception of himself' is reflected in norms which dictate 'where he should be'. It is especially the 'space' of the seaman which has intrigued me of late. It is for that reason that I would like to touch further on the various areas on the vessel itself.

What factors affect the areas of the vessel - and how? It would appear that they can be divided into three groups: 1) the effect of the route 2) the effect of measures taken by the ship's owners, and 3) the effect of other factors, notably governments and trade unions.

## 1. The Route

Route character plays an important role in determining the vessel's size as well as the disposition of its space. Is the vessel to be used, for example, in coastal traffic or on ocean routes; will it be transporting dry goods or liquid cargo, or both at the same time? Is cold storage required? Is the vessel intended to run in the tramp trade or in regular liner shipping? The questions could go on and on.

## 2. The Shipowner

The shipowner's, that is, the investor's, profitability requirements dictates the division between income-producing space (cargo) on board and non-income-producing space such as the crew's living and social quarters. The point is, of course, to maximise the income-producing space and to minimise the rest. The demands of the shipowner must be met by the builder's technical know-how, with the technical prereq-uisites that exist at the time of the vessel's construction.

## 3.   Measures Required by Central Authority and Labour Unions

The character of the vessel and its spaces are also affected by governmental laws and other measures, as to the appropriate amount of space to be provided for particular personnel, dependent on the size and business and voyaging of the vessel concerned.   Moreover, con-siderations as to 'spacing' are also influenced by 'labour-political' bargaining strengths, or lobbying by organised seamen's labour on governments and the influencing of their maritime law making.

Labour union interest in living conditions is easy to understand, and great advances have been achieved.   In the 1930s, the three-masted barque, *Sigyn*, in the Baltic and West European trades, had its five seamen sharing a foc'sle with a floor area of 11.5 square metres. While on the *Winden* in 1991 each sailor had a single cabin with its own shower.

The Finnish laws on living space have undergone striking change.   In 1939 the Finnish law on mercantile shipping did not even mention the matter of living space on board vessels, and not until 1948 do provisions first appear.   In a statute of that year it was laid down that sleeping cabins on ships with a capacity of over 300 tons had to be located over the cargo water line *(lastvattenlinje)* in the middle of the ship or, alternatively, astern (in other words, not at the bow).   It had previously been usual to house the crew under the forecastle at the bow, which for many had proved fatal when vessels were in collision.[9]

On the other hand, the law of 1976, required that the living cabins should have access to daylight, whilst the area of a single crew mem-ber's cabin must, for example, on a vessel of over 3000 tons (such as *Winden*) be at least 4.25 square metres.   The law also contained a detailed description of what must be found in such a cabin.   Besides a bunk, the cabin had to have:

> A ventilated, lockable wardrobe with a shelf, as well as suitable places for the storage of clothing....   The inner dimensions of the wardrobe must be at least 1.8 x 0.6 x 0.5m.   There must be a mirror placed in every sleeping cabin, cupboard for personal items, a bookshelf, and a suffi-cient number of clothes hooks.   Windows must also have suitable curtains.[10]

Despite the fact that the living quarters described in the most recent law are by no means luxurious, at least in the case of sea-hands, they do contain the most necessary requisites for well-being and for a 'normal' life as compared to conditions on land. Developments in this area over the last half century have been tremendous to say the least. The relative living space enjoyed by the captain, officers and crew, the degree of comfort, and their location in relation to each other, again are significant. The sleeping cabin, its location and fittings are, naturally, important for seamen, and form part of the seaman's thought world and affect his conception of himself. Similarly the places where the seaman carries out his daily work also affect the seaman's conception of himself and his place on board. Whether he works on the bridge or in the machine room, on the deck or in the galley, has its significance. It is noteworthy that the different categories of workers use and move about in different parts of the same vessel during the course of their working day. It is self-evident that the bridge and machine room, for example, are, indeed, two very different 'worlds', as are the deck and the galley, for example. It is therefore my intention to map out *where* on the ship each crew member moves. This could later be developed to encompass where they go when in port, how they live at home (on land), and so on. All of this is reflected in the seaman's thought world and helps to form his conception of himself.

## The 'Status' of the Seaman

Here we find a natural point of transition to the third and for us central term, that is, 'status', as used in a broader sense. According to Weibust, status can be 'broadly defined as the relative position within a hierarchy'. Thus, he continues, 'Status may be measured in terms of: (1) subjective status (how the subject views his own position); (2) objective status (on the basis of criteria selected by the observer); (3) or accorded status (the position assigned to the individual by other members of the group). Accorded status, which is the outcome of the group's differentiated perception of its members and leads to a set of status rights and status obligations is, according to Weibust 'of greatest interest to ethnologists'.[11] This may indeed be the case, but for my purpose I am more interested in the first-named status type, that is 'the subjective status, how the subject views his own position'. I would like to quote two passages from Finnish maritime fiction written in the 1960s:

> I was then the furnace tender, the nobility of the boiler room. The hands were toiling under the sharp whip of my fast shovel as I threw my coal into the flames. Oh, endless coal![12]

I had the feeling that I was witnessing the clash of two uncompromising worldviews: on the one hand there was the captain whose every cell oozed with the pathology of a dictator; then there was the independent-minded sailor whose integrity remained intact, in whose Finnish heritage it was to never blindly obey orders.[13]

It is utterances such as these that best reflect 'the seafarer's conception of himself'. This is precisely what, as a researcher, I am interested in.

## Other Perspectives

Status on board can also be discussed from a number of other and different perspectives, of which I would like to mention a seaman's 'personal qualities'. These would include his skill, his knowledge and experience, and his possessions, that is, things belonging to the individual and especially items which would typically belong to people who work at sea. Among these possessions, I will only name *the broad-bladed sheath knife* and its colourful history. Given what I personally have witnessed, it still seems today to be one of the distinguishing features of a seaman.

## Problems

I will conclude these preliminary observations by briefly touching on the most obvious problems of the study in mind.

I have chosen to consider the seaman working on dry cargo vessels, on routes in the North Sea and the Baltic, and in the three eras of the sailing ship, the steamer, and the modern motor vessel. The study thus spans three time periods, between which it is intended comparisons will be made. But we need to keep in mind the fact that source material from each of the different periods is very heterogeneous.

From the era of the sailing ships in Finland, there is archival material in the form of manuscript material, photographs, ships' plans, later scientific studies, fiction and recorded interviews. In addition there are a number of preserved, museum ships. If, on the other hand, the comparison is made, for example, with a 1950s freighter, there are fewer interviews, less fiction and no Finnish museum ships, to use for such a purpose. In the case of modern shipping there is, on the other hand, 'too much' material at the disposal of researchers, while too little of it has been worked through.

One needs also to ask, at this early stage of the study, whether the modern ship, the *Winden*, I have chosen to study, and its crew can be seen as representative of modern shipping in general. In fact it is clear no one vessel or type of vessel, engaged in its particular business, can be seen as representative of modern shipping and the great variety of business that is undertaken. So comparison will need to be made of the *Winden* and its crew with that of other categories of vessels and their crews.

## NOTES

1   This project is under the general supervision of Professor Nils Stora of the Åbo Akademi University.
2   Marvin Harris, *The Rise of Anthropological Theory. A History of Theories of Culture* (London, 1968), 571. See also, K. Pike, *Language in Relation to a Unified Theory of the Structure of Human Behaviour*, Vol. 1 (Glendale: Summer Institute of Linguistics, 1954).
3   Knut Weibust, *Deep Sea Sailors. A Study in Maritime Ethnology*. Nordiska Museets Handlingar, 71 (Stockholm, 1976).
4   Cited on the back cover of, Uno Salminen, *Till Sjöss* (Jakobstad, Sweden, 1977).
5   *Ibid*.
6   Niilo Wälläri, *Antoisia Vuosia. Muistelmia Toiminnasta Ammattiyhdistysliikkeessä* (Helsinki, 1967), 166.
7   Satu Lassila, *Puhuminen Tyynnyttää Mielen Myrskyjä*, Vapaavahti/Frivakt, 9-10/91 (Helsinki, 1991), 10.
8   Richard Henry Dana, Jr., *Two Years before the Mast. A Personal Narrative of Life at Sea* (Glasgow, 1st ed., 1840), 10.
9   *Finlands författningssamling No. 794* (Helsingfors, 1948).
10  *Författning om besättningens bostadsutrymmen ombord på fartyg*, 17.6.1976/518. Finlands lag 1992, År 13 (Helsinki, 1992).
11  Weibust, *Deep Sea Sailors*, 258.
12  Eino Koivistoinen, *Sininen Meri* (Porvoo, 1967), 40.
13  *Ibid*, 110.

Kim Montin was keen to point out when giving his paper at Darting-ton in 1992 that the first time he had become acquainted with this 'beautiful country of yours', and Devon, was twenty-six years before, when a young student at the Plymouth Sailing School. He added that he would never forget the fascinating sailing along the south coast of England, visiting the Channel Islands and St Malo, nor the 'weeks of sunshine and good winds'.

# OCCUPATIONAL CONDITIONING: FIRST VOYAGE IN *PASSAT*

## Alston Kennerley

This paper investigates the decision by the author, in 1951 and at the age of 16, to adopt a seafaring career and to undertake his first voyage at sea, which unusually for the period was in sail. Whether, forty years on from these events, it is possible to step aside from personal experiences and attempt an assessment which is not completely subjective must surely be questionable. The limitations of such an evaluation include the difficulty of achieving an acceptable level of objectivity, and the necessity of acknowledging and revealing personal factors which would otherwise remain hidden. A context for the autobiographical sections of the paper is provided by considering theoretical aspects of industrial socialisation and by drawing on other sources illustrative of seafarers' backgrounds and initial experiences.

The term 'socialisation' needs some explanation as a preliminary to the main discussion. The process of socialisation is the process of learning to live in society.[1] It starts at birth and continues throughout life, and involves the transmission of particular cultures. It can also mean the process of becoming human in which attributes unique to an individual are developed. Major influences in the process are parents, family, friends, peers, school, teachers, the physical environment, the media and the employment environment. At major changes in a person's life, new layers of socialisation are added. Earlier layers may lapse, but gradually a person learns to live in several different social contexts adapting behaviour when changing from one to another. Extensive previous learning makes it easier for a person to adapt to new social contexts.

The socialising influences of the workplace often constitute a particularly strong form of socialisation, involving particular behavioural patterns associated with the learning and performance of technological skills. Indeed the process is reinforced by the reward system: failure to behave and perform satisfactorily may lead to loss of employment and thus income. In seafaring, these elements are further reinforced by the closed or 'total' nature of the society. Ships are grouped with prisons, monasteries, barracks and boarding schools, in the way and degree to which they subject the members of their societies to an all pervading or 'total' environment excluding completely other social contexts in which the mass of people outside (ashore) move simultaneously.[2]

Finally, while all seafarers are subject to the special environment noted above, there are several seafaring careers each with its own sub-culture. The apprentice/deck officer (with command as the goal) in the British Merchant Navy was the career upon which the author embarked, and the one principally represented in this paper. Parallel officer careers included those of marine engineer and radio operators. Ratings careers included those of deck hands, stewards and engine room hands. Different career structures for similar tasks at sea have been provided in the Royal Navy.

## Anticipatory Socialization and Career Choice

It has long been the case that the typical merchant seafarer embarked on his career at about the age of sixteen, and in recent times that has been at the completion of secondary education.[3] Thus the choice has been conditioned by the awareness gained up to that age of forms of employment with which a young person has had passing contact. While the childish desire to become an engine or bus driver is well known, it has long been recognised that parents', especially fathers', careers are a strong influence. This is noticeably so among seafarers.

In an analysis of the youngsters entering the Boys' Department of the South Shields Marine School between 1886 and 1900 it was found that about half the boys could be described as coming from seafaring families (see Appendix 1). Of the 194 boys for whom the father's occupation was recorded, 67 (34 per cent) were master mariners, 14 (7 per cent) were marine engineers and 16 (8 per cent) were in marine-related occupations such as rigger or ship manager. Of the remainder 17 (9 per cent) were noted as engineers (likely to be working in the marine sphere) while the balance of 80 (41 per cent) came from a wide variety of occupations as diverse as cartman, insurance agent and min-ister. A similar balance was found in a study undertaken in 1971 where again 50 per cent of the sample came from family backgrounds having seafaring connections. In the case of deck cadets those connec-tions were with ships' masters or deck officers while engineering cadets had family connections with engineers.[4] The pattern of seafarers' sons following their fathers was noted in a social survey in Liverpool in the 1930s. They were more likely to do so than in any of the other eight occupations examined.[5]

Living in a seafaring community was also of great influence lead-ing to the choice of a career at sea. Again there is ample evidence. Tony Lane shows in his research among Liverpool seamen how in the 1960s as in the 1930s living in Liverpool's seafaring community close to the docks and the merchant ships using them, being in contact with the continual arrival and departure of seemingly wealthy seamen from and to exciting and exotic places overseas, being regaled with stories of exploits abroad, led to the assumption of a seagoing career without consideration of alternatives.[6]   Support for this influence comes likewise from the research of the Gwynnedd school of maritime history led by Aled Eames, in his own publications and in those of his associates, as well as in articles in *Maritime Wales*. Similar sentiments are to be found in numerous seafaring autobiographies, including Captain C. Fenton in *The Sea Apprentice*, and A.H. Rasmussen, who uses the title *Sea Fever* to evoke a lifelong fascination with the sea.[7]

While Liverpool's seafaring community provided an environment of seafarers on leave, it was not nearly so easy for youngsters to spend much time on board ships and boats in port, compared with the communities in the smaller ports in other parts of Britain (such as those in North Wales). Autobiographies often note playing on board ships, short voyages as children or simply 'messing about in boats' as factors in going to sea. Weibust identifies this as a process called 'anticipatory socialisation' in which preparation for a life at sea starts on shore thus easing the adjustment when seafaring proper commences.[8]

Despite this strong evidence of continuity between generations of seafarers, by no means all seafarers have come to this career through such influences. On shore it has not been unusual for young adults to experiment with several occupations during their early working years before settling to a particular career.[9]   Frank Bullen scraped a precarious existence as a 'street Arab' with a variety of short-lived jobs for three years in the Paddington area of London, having some contact with seafarers. He says he was well aware of the unpleasant rigours of seafaring, yet, aged nearly twelve, he persuaded an uncle to take him to sea as a cabin boy.[10] J.F. Ruthven belonged to landed gentry in Ireland, and would have made a career in the army except that family finances were inadequate for supporting the lifestyle. So he selected the sea and trained on H.M.S. *Conway* (1864-5).[11] Others, like Geoffrey Rawson, were ordered to a career at sea by their parents.[12]

Although trial voyages have by no means been rare, such experimentation has not been easy owing to separation from the home employment market and the need to accept unemployment while changing jobs.  For the author's generation of seafarers, testing other careers was made virtually impossible by the National Service regulations which required service in protected occupations (such as mining and the Merchant Navy) up to the age of 26, if two years National Service after the age of 18 was to be avoided. A further restriction for deck apprentices was their indentures which bound them legally to their employer for up to four years.

Research in the past twenty years into seafarers' motives in select-
ing careers at sea, has shown other factors at work.[13] Those experiencing difficulty adjusting to a wider society, are attracted to the relative isolation of the ship. Persons coming from remote socially-
deprived communities are attracted to the relatively fuller shipboard
life, and have been found to be more likely to remain in a sea career
permanently. It has also been shown that a higher proportion of persons from broken homes are to be found in seafaring careers than in
shore occupations. 'Relative deprivation', the term under which such
factors are grouped, is also an element in wastage from sea careers,
which has always been high compared with other occupations. Here,
often at marriage, the seafarer recognises the fuller social context
ashore which he has missed during his life at sea.[14]

For almost all the examples quoted above, the most formative
years for an awareness of career possibilities were probably between
the ages of ten and sixteen. But for the author, these years were spent
in comparative isolation from contact with seafaring influences. His
parents had separated and divorced, his father passing out of his life.
He was sent to boarding school, initially in Southport, but from twelve
in rural Shropshire, and holidays were spent with his mother who had
moved to an inland village in North Wales. Visits to Liverpool, where
his grandmother lived, were rare. He has no recollection of giving any
thought to the sea during this period. He was, however, aware that the
cost of keeping him at boarding school was a severe drain on family
finances.

As was common at that time, his school provided none of the
career awareness classes that are so common now, thus his career
examples constituted his teachers, the school chaplain, and farmers.
Those pupils that remained after sixteen were being prepared for
university entrance, and this constituted the only form of progression
for which the school provided any understanding. The author reached
the age of sixteen without any idea of a career and with no
understanding of the significance of the 'O' level examinations he was
taking or of the educational ladder.

In the summer of 1951, his mother exerted pressure to choose a
career, obtaining literature on a variety of occupations of which one
was forestry. None offered any strong appeal, and without any deep
thought, he decided on a career at sea. A single letter to Alfred Holt &
Company (the Blue Funnel Line), and an interview in Liverpool, set
his path for the next ten years.

Given the meagre contact with ships and seafaring over the past
six years, how was it that the author drifted so casually and easily into
this career? The answer would seem to lie in socialising influences
before the age of ten. There were three elements to the family dimen-

sion. For some twelve years his mother had worked as a clerk in Alfred Holt's headquarters in Liverpool, India Buildings, and was fond of regaling her experiences, which included a reward holiday voyage to Egypt. His father trained in the mid-1930s as a Radio Officer sponsored by Holts and went on to serve in their ships throughout World War II. Although mostly an absentee father, it was inevitable that his ships and voyages featured in life at home. Thirdly, his mother, who was herself the product of a broken home and an austere mother, re-established contact with her father, Captain W.G. Wainwright, and in visits to him made as a four-year-old, he was always referred to as 'The Captain'.[15]

The author experienced additional 'anticipatory socialisation' through his father being transferred in 1939 to the M.V. *Charon*, a Blue Funnel ship based permanently in Australian waters.[16] This led in November 1940, to he and his mother being sent to reside in Fremantle, Western Australia, aboard the S.S. *Ulysses*. Of this voyage as a five-year-old, he remembers ship's gunnery practice, being rescued from the forward well deck by a seaman, and the transfer of missionaries to surfboats by 'mammy chairs' during a call at Lagos, West Africa. His mother has recorded:

> [whilst loading in Takoradi (West Africa)] I lost my young son there and eventually found him. He was leaning over a rail where there was an open hold, holding a long piece of string. The natives were up and down on ladders laden with cocoa beans. I said what are you doing. He said 'fishing for men' (shades of Sunday school). There was much laughter and chatter. It appeared that he was tying a toffee to the end of the string and lowering it into the hold one at a time.

Whilst living in Fremantle, these experiences were sustained by visits to his father's ship when it was in port, and through his father's shipmates visiting the family home. Other influences included a visit to H.M.S. *London*. The return voyage to Liverpool aboard Holt's S.S. *Nestor*, came in the spring of 1945, before the war ended. On this passage the whole family were together, the author's father serving as auxiliary radio officer.

In retrospect, it is clear that this dormant socialisation played a significant part in the author's decision to go to sea, particularly in the absence of any strong alternative example. He was subconsciously aware not only of the geography and function of merchant ships, but also of the behavioural pattens required of seafarers, and of the hierarchical distinctions between the different sub-groups aboard ship. Under pressure to make a decision he settled for the familiar.

## Outward Bound Training

For most boy seafarers socialisation for a working life aboard ship, apart from signing articles in the shipping office, only started when they set foot aboard their first ship, usually just before it sailed.[17] But with apprentices this was not always the case. Once indentures were signed, the shipowner became responsible for their maintenance. Often the ship was not ready to sail. In some cases boys were placed on board under the oversight only of the ship keeper; some were sent home; in other cases they were accommodated with a navigation teacher and taught seamanship and navigation. Captain Fenton was placed in the care of Captains Cogle and Taylor who ran a Nautical Academy in Upper Pitt Street, Liverpool.[18] Several boys attending the South Shields Marine School in the 1890s, in the example above, fell clearly into this category. Indeed, nautical schools have long been in existence which provided 'preparing for sea' courses.[19]

This form of 'anticipatory socialisation' was developed during and after World War II, particularly for Holt's personnel, through the influence of Kurt Hahn, the progressive German educationalist, who explored his ideas on self-sufficiency and 'health giving life' initially at his school at Salem, Germany, and then at Gordonstoun in Scotland.[20] These ideas influenced Lawrence Holt, and with his company's financial support, the Outward Bound Sea School was established at Aberdovey in Wales in 1941. Holt's main purpose was to provide basic seamanship training for seafarers who rarely had small boat experience yet were at that time increasingly likely to have to take to ships' lifeboats following a submarine attack. He was not, however, averse to Hahn's concept of a training through the sea which could benefit all walks of life. The School was staffed in its early years by Holt's employees, all seafarers, and thereafter all new Blue Funnel entrants spent a month on an Outward Bound course before joining their first ship. In this way all Holt's apprentices came to be subjected to a form of 'anticipatory socialisation' prior to joining their first ships.

The author's experience was to be no different, except that along with nine other new apprentices he had been offered an opportunity by Holt's to serve on one of the two sail training ships being fitted out in Kiel, in 1951, the four-masted barques *Pamir* and *Passat*.[21] The influence of Kurt Hahn was present in this development as he saw the mixing of cadets on the ships as a means of breaking down barriers between Britain and Germany following the conflict. In the end the author was to experience one of the longest ever outward bound courses. In the autumn of 1951 he was one of five British apprentices selected to undergo a special outward-bound course at Kurt Hahn's

Gordonstoun to which about fifteen German cadets had also been invited. All the cadets, in excess of one hundred, were assembled for further outward-bound training, in December 1951, at Schloss Nehmten, a mansion on the Ploner See near Kiel, where they remained until the ships were ready. The author was not able to join his ship until February 1952.

At Gordonstoun, all Hahn's methods of physical and character development were put into practice. The day started with an early morning run followed by a cold bath. Personal attainment targets were set for a range of physical activities, including swimming and running, orienteering and the obstacle course. Practical seamanship included ropework and rowing and sailing in cutters at nearby Hopeman Harbour, augmented by a trip in the schooner *Prince Louis*. Food was wholesome, and included the novelty of homemade muesli at breakfast. There were the usual domestic duties to perform. Transport was by bicycle. Evenings were filled with activities such as Scottish dancing and learning sea shanties, and on one awesome occasion the British apprentices attended a formal dinner with Kurt Hahn. There was little spare time.

After a few days at home and at Holt's office in Liverpool, the British apprentices travelled by rail to Germany, to join a much larger 'Outward Bound' course. The seafaring socialisation was stepped up by being organised in watches, through mounting night watches on the ship's life boats where rowing was practised, by classes on the rigging of square-rigged ships and by the presence of some of the ships' officers as temporary instructors. Many of the activities experienced at Gordonstoun were continued, including early morning runs, the obstacle course, long spells rowing on the Ploner See. Other classes included navigation and first aid. Domestic duties were undertaken by each watch in turn, comprising general cleaning, carrying wood for the stoves and boilers, vegetable preparation (especially mounds of potatoes), serving at table and washing up. The performance of each watch was monitored by a points system. Recreation included volleyball, and evening events included general interest talks and parties. Stan Hugill, the authority on sea shanties, was brought over from Aberdovey to teach sea shanties, the British cadets taking the opportunity to note down some 'authentic' words.[22]

The establishment was being run on a pattern which owed something to that practised on German sail training ships before World War II, but which would not have been unrecognisable aboard British training ships, such as *Conway* or *Worcester*, which were still operating at that time. However, while the British contingent could relate easily to the Outward Bound elements and were able to cope with the discipline and the practical classes on ship's rigging, they had to adjust to an added layer of socialisation, the German culture and language. A limited amount of informal language instruction was provided, but in the main the British had to pick up the basics for survival as best they could. The language was learned aurally, and although thinking in German did eventually develop, the author's vocabulary was limited to shipboard contexts. Adjusting to German social practices came fairly easily, though some of the food was never appreciated.

As the *Passat* neared the end of her refit, the cadet body was taken to Kiel to provide the labour for taking her sails back on board from store, and they were on board for a day during trials at sea. But she was sailed round Denmark to Brake, a small port on the River Weser, under her crew of experienced seafarers only, the cadets joining her there on 31 January 1952. Of the original ten Holt's apprentices, two had been withdrawn, four had been selected to join the *Pamir*, leaving the author and three colleagues to serve in the *Passat*. It must be emphasised that such an extended period of 'anticipatory socialisation' and training, was then unusual, unless pre-sea courses are included in the definition. Many British apprentices joined their first power-driven ships with minimal preparation.

**Sealife in Sail**

Although the cadet body which joined the *Passat* had lived together in boarding school conditions for over a month after the departure of those aboard the *Pamir*, and had thus developed social relationships between each other, these were forged in the context of the social regime imposed at Schloss Nehmten. The existence of these relationships, the instruction about the ships, and the advance visits certainly eased the arrival on board; none of the cadets could have felt the gnawing anxiety and isolation experienced by the lone first voyage apprentice as suggested in some autobiographies where careers started in sail. All were 'first trippers': there was safety in numbers and implied mutual support. Nevertheless, there was now a team of officers and crew, all of whom, it seemed, had some degree of authority by virtue of rank or previous seagoing experience. There were new social relationships to learn, as well as the need to adjust to ships' watchkeeping arrangements and work stations, which broke down previously established relationships. Thus, this fresh interpersonal and job socialisation was by no means avoided by the pre-sea stage.

In theoretical models of organisations, the individual learns a basic level of technical competence, and then is subjected to pressures to conform to the organisation and pressures to improve technical ability which will allow advancement.[23] The individual modifies his behaviour to match his position in the organisation and to match the technical requirements of the job. Conflict is created if this is not achieved sufficiently quickly, or if the individual attempts to move outside his perceived role. All organisations demand a degree of conformity and subordination of the self to their requirements from their members. In some cases, for example where processes depend on manual dexterity, the organisation, through requiring near robotic performance, may demand total subordination of individuality. The more extreme total institutions, such as prisons, require subordination of individuality to achieve absolute conformity as a technique of social control.

Ships at sea are pieces of machinery on the move. The object of shipboard organisational structures has always been to get the ship and its cargo to its destination in safety. Thus it is important that individuality is subordinated to this operational requirement, and that the various tasks that achieve this are carried out in the standard manner necessary. On a sailing ship, this is a continuous activity, for which manpower is arranged on a shift basis. When on watch, its crew works within the propulsion unit, adjusting and optimising the moving parts to achieve the best propulsive output. In control system terms, the work of the crew is the switching mechanism and the lubricant, responding in detail to the control decisions (orders) of the officer-in-charge. It is important that novices learn early that rapid and accurate responses are required to this highest-level demand on them, and that all other ship-imposed or personal activities are immediately abandoned when the orders are given.

Off duty, there might be more scope for individuality, if any time remained after sleeping, eating and domestic duties. On power-driven vessels the normal four hours on and and eight hours off watch routine, and the allocation of 'day work' when some crew were excused watches and worked on maintenance duties by day, gave some scope for time which the individual controlled. Even so, the limited spatial environment and the proximity of other members of the crew, meant strict conformity with standard shipboard social structures. The opportunity for even this limited individuality, was often non-existent on a sailing ship, where all except a very few 'idlers' kept 'double watches' or 'watch and watch about'. Working, eating and sleeping, with the occasional 'all hands on deck' when off watch, left little time for anything else: the environment became total through the operational demands of a ship under way.

Compared with the manning levels which had obtained when the *Passat* was owned by Gustaf Erikson in the 1930s, the combined cadet body, officers and seamen, was three times as large in 1952. Nevertheless, double watches were the norm and the demands for conformity were by no means lessened. There was, of course, a difference, in that

*Plate 1.* The four-masted barque *Passat*, 1952, in the North Atlantic.

*Plate 2.* The author on the fo'c's'le of *Passat*, 1952.

the deck crew were just sufficiently numerous to be able to handle the ship, had the cadets not been aboard. The fifty or so first trip cadets, were certainly a liability, despite their pre-sea training, until they had internalised their assigned stations and literally 'knew the ropes'. The additional manpower was an advantage from the start if muscle could be 'laid on' to the right rope quickly enough under the direction of an AB.

*Passat's* cadets understood the need for rapid response from the outset, although by no means all managed to learn the ropes early enough in the passage. In many manoeuvres (going about, setting, trimming or taking in sail) the abundance of manpower at any particular station usually ensured that one of the group knew what to do. Verbal hazing certainly occurred, but as often as not it was directed at the cadet body as a whole rather than at individuals. Certainly the kind of bullied learning (verbal abuse and physical punishment) which Weibust analyses in his research into sailing ship crews, was hardly evident.[24] Most individual punishment was of a school nature and associated with slackness and misbehaviour.

Night watches in a steady wind, often meant being on deck with nothing to do until turns at the wheel or lookout were due. Being caught asleep on a hatch or sheltering in an alleyway, led to being kept on duty when the rest of the watch had been relieved. Arriving late for the watch muster produced a similar effect, but more effective was the disapprobation of the watch kept waiting for its relief.After several weeks, one cadet's lack of knowledge of the ship's rigging was exposed. He was rewarded with the public and physically arduous punishment of climbing each mast in turn, working his way to each yard arm from which he was required to hail the deck naming the rigging. More often misdemeanours were punished *en masse*. On one occasion a whole watch was required to run several times around the decks from stem to stern. 'Field days' were a standard method of group punishment: ship's work carried out during a watch below. This was effectively a loss of privilege; its significance as a punishment was undermined because 'field days' were often set simply to complete maintenance work, for example, before arrival in port. One unpleasant task which had to be undertaken, whether as a punishment or not, was breaking up very large lumps of coke for the galley stove. The coke was stored in the lowest level of the fore peak. As the supply of beer for the officers was stored in the level above, the beer was of course raided.

Handling sail from the deck was nearly always associated with work aloft. Sails to be set had to be loosed from their gaskets; sails taken in had to be furled. Invariably at the end of a sail manoeuvre, an anti-chafing proceedure, overhauling the buntlines, had to be performed aloft, while on deck the numerous ropes handled had to be coiled up or flaked down ready for the next manoeuvre. These post-manoeuvre jobs fell to individuals, mostly cadets once they were reliable. Certainly, returning to deck after completing even these elementary tasks aloft, produced especially at the beginning, a feeling of elation at a job completed in difficult circumstances (as experienced, for example, by mountaineers).

Climbing and working aloft certainly demanded agility, particularly out on yard arms and near the top of the masts where footholds and handholds tended to disappear. Experience aloft developed the delicate balance between relaxed movement and maintaining personal safety, and also consideration for others aloft as one person's movements usually affected persons nearby (as when standing on foot ropes furling sail). Initially, *Passat* cadets were required to wear a safety lanyard aloft, a short length of rope made fast around the waist, with a clip hook spliced to its tail end. Though it helped to overcome initial fears, movement was hindered by being clipped to say a jack rail on a yard. They were soon ignored partly for practical reasons, but more perhaps, because their use was the mark of a novice. Tasks, such as reeving flag halliards through the sheeve in the mast cap were particularly awkward owing to the lack of holds. The author always feared, though was never asked, the job of clearing the ship's name pennant on the jigger mast, which had a lengthy section of bare mast to be climbed.

Two other routine watch-keeping tasks were steering and lookout. Cadets at the wheel were at first supervised by a seaman, but thereafter they were alone unless heavy weather demanded extra hands to control the wheel. Mostly the helmsman could hold a helm position by standing on the friction brake. With the sails well balanced very little movement of the wheel was needed to maintain direction, the author on occasion being able to achieve this with out any wheel movement for ten minutes at a time. The helmsman was also the ship's time keeper and had to remember to strike the bells on the half hour. Lookouts were posted on the fo'c's'le head at night and in fog. In answer to the bells struck by the helmsman the look out repeated the pattern on the fo'c's'le bell, then, using a large megaphone, hailed the officer of the watch with the cry '*Auf die Back ist alles wohl, die Lampen brennen hell und klar*'. Doing this on the first occasion was a minor milestone for the author. In fog, the lookout was subjected to the intense vibration caused by the electric fog horn, mounted by the bell at waist height.

Progressively the whole range of maintenance tasks devolved on the cadets, though often more important work on the rigging was undertaken as assistants to one of the seamen, the sailmaker or the carpenter. Chipping rust, painting and scrubbing decks were mass activities which in fair weather continued throughout the day, without a break as the watch on deck continued with the task set (as on lookout or at the wheel) until relieved after the new watch had mustered. Work aloft was different. Usually the job in hand was completed even if it meant losing part of a person's watch below.

To those maintenance jobs which could be done at sea, were added in port work which could not be done underway, including chipping and painting overside and work in the holds. The outward-bound cargo of bagged cement left two feet of loose cement in the holds which had to be bagged before the holds could be cleaned for the homeward cargo. This included grain in bulk for which shifting boards, a temporary wall along the centreline, had to be rigged. Homeward the ship sailed trimmed by the head, which led to a quantity of bagged grain being moved over the deck into an after hold. Other work in port included managing the large crowds which visited the ship on open days, for which the cadets mounted gangway duty and provided guided tours.

Though the cadets worked as part of the team, their time off duty was generally passed separately from the crew and of course from the officers, a feature dictated in part by the accommodation arrangements. In order to bring the ship up to date and to enable her to carry some fifty cadets, the general arrangement of the ship was extensively modified during the refit in 1951. Cadet accommodation was provided at 'tween deck level at the stern of the ship, comprising port and starboard mess/hammock rooms and a small communal dayroom in the way of the counter stern. Their showers and toilets were in the poop, which otherwise accommodated the seamen, space which the cadets did not enter. The officers and petty officers lived in the midships house, the forward end of which contained the galley. Again this was space which was not normally entered, except to fetch food from the galley. On deck, the fo'c's'le was communal space, and here, as on sailing ships from time immemorial, cadets continued their socialisation by listening to the yarns of the older seamen, especially those having square rig experience between the wars.

The cadet mess rooms each housed about 25 cadets. Storage for personal effects was in lockers along the centre line bulkhead, with a full height and a half height narrow locker allocated to each cadet. A basic uniform clothing outfit had been provided by the the German owner before joining, and the cadets had made their own kitbags and hammocks.[25] Portable mess tables and benches were erected during the day, but folded into deckhead racks by night. Cadets slept in hammocks provided with thin narrow mattresses and a tiny pillow. Only about two thirds of the hammocks could sling at the same level, so some slung high and others low. The author always slung high, stretching his hammock bar tight up to the beams and sleeping on his stomach. Stretchers were frowned upon as they interfered with the space of others. By day the hammocks were furled and stored in side benches. Sheets and a blanket completed the bedding. As the ship was provided with a washing machine, changes were provided from time to time, and top clothing could be washed communally, the cadets providing the labour. Inspections of clothing took place on occasion, particularly approaching port so that a reasonably smart uniform turnout was available for public consumption.

Food, in time-honoured form, had to come over the midship house and aft to the cadets' quarters. Cadets took turns at '*Backschaf*' (peggy or mess steward) fetching and serving meals, washing up and cleaning. The fitting of increased steward's storage, including refrigerated space as well as dry storage, meant a more varied diet than could have been managed in the past. Despite increased water tank capacity, the large crew meant the supply had to be managed with care. Water was rationed, and in the doldrums a rain tent was rigged to augment the supply, while the opportunity was taken for rain baths on deck. A longer than anticipated passage homeward left only pea soup for the last few days.

As well as the *Passat*'s master and deck officers, the saloon comprised the engineers responsible for the auxiliary engine, generator, refrigeration plant and pumps, a radio operator, meteorologist and doctor. All were drawn upon from time to time to continue the educational programme, though classes only occurred on deck during fine weather. The doctor was provided with a surgery and sick bay in the after deck house. Mostly he dealt with minor complaints, though on one occasion he operated for appendicitis, assisted by the second mate and the baker. *Passat* was considered a lucky ship as there were no serious accidents to personnel, and she encountered very little serious weather. A tornado when berthed in Rio Grande do Sul in Brazil, left the ship swinging out into the harbour with five broken head ropes and a fo'c's'le mooring bollard snapped off.

The scale of the manpower available, made it possible for double watches to be broken whilst in the trade wind zones. The three-watch-system which was adopted provided more time for the classes noted above and for personal free time. Conditions permitting, Sundays were freed of all but essential watchkeeping duties, and were often marked by the provision of some delicacy from the galley. Sometimes records were played over the ship's loudspeaker system. In port, special trips away were arranged. The German communities in the South American ports received the German cadets, on one occasion for a three day visit, while English families played host to the British cadets.

Regular leave ashore was allowed in port, and in Rosario and Buenos Aires in Argentina the British cadets found a ready welcome in the Missions to Seamen clubs.

## Cultural Integration and Conclusion

Despite the lengthy period of socialisation the British cadets still retained much of their earlier layers of socialisation, which led at times to differing responses to situations from the rest of the cadet body. Though the German cadets were also first trippers, they were not apprentices; they were boy seamen and had to pass through the seaman ranks before even considering studying for officers' licences. The British cadets knew that at the end of their apprenticeships they would take their first professional certificate and become deck officers. The Germans received no pay except ten marks per month pocket money. The British were in receipt of about six times that amount at home, though for their time in *Passat* they were allowed only the same as the German cadets. Thus physically there was no distinction, but their career culture differed. The author kept up navigation studies when off duty and with his colleagues pursued the seamen's hobbies of fancy canvas and knotwork with the aid of the *Ashley Book of Knots* borrowed from one of the seamen.[26]

Although the four British apprentices on board *Passat* were paired in different watches, they remained a sub-group slightly apart from the Germans. Their limited German vocabulary meant that they missed the subtleties of conversations amongst the German cadets, and thus were perhaps insulated from absorbing the culture fully. This insulation probably meant that they were not party to some of the social stresses which developed amongst the Germans, until they emerged in violent argument. Certainly they felt at times that some of their colleagues lacked enthusiasm and even initiative. They were fully conscious of the special nature of the opportunity service in the ship offered, and were keen to maximise their square-rig experience, by undertaking as wide a variety of sailorising jobs as possible. Thus they welcomed all opportunities to understudy experienced seamen and at times undertook interesting tasks in their watch below. But like their German shipmates, they were impatient with the extent to which the auxiliary engine was used when winds were adverse, and more than pleased when a propeller blade broke off preventing its use.

It is of course possible that they were treated with some care and even favoured from time to time because they were British. World War II was still quite recent. *Passat*'s officers had served in the war, one in a 'U' boat. A number of the German cadets had been in the Hitler Youth. It was known that the English apprentices were required to submit voyage reports to Holts at regular intervals. In fact the war rarely surfaced in conversation. The British reserve probably showed from time to time, particularly when one of the officers attempted to probe the content of their reports and their intentions at the end of the voyage.

In fact after the first voyage of about five months, the British wanted nothing better than to continue in the ship. They had established their place in that environment, and wanted to become 'substantial' square-rig men. Thus on arrival in Antwerp, they sought out the Blue Funnel agent, to get their request to Holts transmitted to Liverpool. This was in contrast to the four who served in the *Pamir*, whose experience had been less successful, and who had left at the end of their first voyage. Holts agreed to a second voyage, and the British cadets were granted a few days home leave. After reporting in person at Holt's office in Liverpool, they rejoined *Passat*, in Bremen, where she was loading outward. But by the end of the second voyage, they had become more concerned about progressing their careers as officer apprentices, and made no attempt to request a third voyage. They left the ship in Antwerp in November 1952.

In retrospect the author had been well prepared for his first year at sea, not only by the seafaring associations of his early childhood. His boarding school experience enabled him to slip easily into life aboard a training ship. In some respects, the fact that it was a German ship was incidental as the industrial socialisation was as a square-rig seaman rather than as a German seaman. There was a degree of irony about his appointment to his next ship, the M.V. *Stentor*. He found himself with a new set of colleagues, and the junior of four Blue Funnel apprentices. There was a new layer of socialisation to be absorbed, particularly that of an apprentices' half deck, where, unlike the equality which pervaded the cadet body on board *Passat*, seniority among the apprentices was measured in seatime to the nearest day. His extensive experience of practical seamanship did, however, mark him out.

## Appendix 1

### Marine School of South Shields Boys' Department

#### Occupations of 194 Boys' Fathers

| Occupation | No. | Occupation | No. | Occupation | No. |
|---|---|---|---|---|---|
| Blacksmith | 2 | Foyboatman | 2 | Policeman | 1 |
| Boilermaker | 1 | Fruit Dealer | 1 | Publican | 1 |
| Bootseller | 1 | Glass Merchant | 1 | Rigger | 1 |
| Brassfounder | 1 | Grocer | 3 | Riveter | 1 |
| Bricklayer | 1 | Harbour Master | 1 | Rope Merchant | 1 |
| Building Insp. | 1 | Head Master | 1 | Sailmaker | 1 |
| Butcher | 1 | Inspector | 1 | Sch.Bd.Officer | 1 |
| Caretaker | 1 | Insurance Agent | 1 | School Master | 1 |
| Carpenter | 3 | Insurance Mangr | 1 | Seamen | 2 |
| Cartman | 1 | Ironmonger | 1 | Shipbroker | 3 |
| Cashier | 3 | Joiner | 1 | Ship Chandler | 1 |
| Chemical Manager | 1 | Manager | 2 | Ship Manager | 1 |
| Chemist | 1 | Master Mariner | 67 | Ship Owner | 1 |
| Clerk | 1 | Mate | 2 | Ship's Husband | 1 |
| Comm.Traveller | 1 | Merchant | 1 | Shipwright | 1 |
| Deceased | 2 | Minister | 4 | Steward | 3 |
| Dock Manager | 2 | Painter | 1 | Storekeeper | 1 |
| Draper | 1 | Pawnbroker | 3 | Tailor | 2 |
| Engineer | 31* | Pilot | 8 | Timber Merch. | 1 |
| Estate Agent | 1 | Plater | 3 | Tugowner | 1 |
| Fireman | 2 | Plumber | 1 | Upholsterer | 1 |
| Fisherman | 1 | | | | |

* Includes 14 Marine Engineers

Source: South Shields Marine School records, kept at South Tyneside College: Admission Book, 1861-1906, Boys' Department Section, 1886-1900.

## Appendix 2

## Alfred Holt & Company

Alfred Holt & Co. were the managers of the Ocean Steamship Company formed in 1865. Holts absorbed the China Mutual Steam Navigation Company in 1902 and in 1935, the Glen and Shire Lines. It was in the ships of these companies that Holts' apprentices, bearing the title 'Midshipman' within the company, served in the 1950s.

Holts had long run their own training scheme for their apprentices, and their association with the German sail training venture must be seen as a manifestation of their continued interest in training, though they did not send any more apprentices to join *Pamir* or *Passat*.

Holt's head office had long been at India Buildings, Water Street, Liverpool, from which seagoing personel were managed. It was its Midshipman's Department which administered the initiative, though the authority undoubtedly came from Mr Lawrence Holt (manager, 1908-1953), who met the group of apprentices before they set off on the adventure. It was to this Department to which the author's voyage reports were sent. His mother worked in the previous building on the site as a clerk for about ten years in the 1920s.

S.S. *Nestor* was a cargo/passenger liner built by Workman Clark, Belfast in 1913. Length 563.2 ft., Breadth 68.4 ft., Depth 40.2 ft.. GRT 14,628. NRT 9,100. Twin-screw triple expansion steam engines, IHP 5,500. She was scrapped in 1951.

S.S. *Ulysses*, China Mutual Steam Navigation Co., GRT 14,646, was the sister ship to the *Nestor*. She was sunk by U160 on 10 April 1942.

M.V. *Charon* was a cargo/passenger liner built by Caledon, Dundee, in 1936. Length 333 ft., Breadth 51 ft., Depth 24 ft. GRT 3,964. NRT 2,246. Official Number 164,308. Six cylinder B & W diesel engine, NHP 644. She was one of three sister ships built between the wars for the trade between Western Australia and Singapore, and was also equipped to carry refrigerated cargo and livestock. The author also served on her, in 1959-60. She was scrapped in the mid-1960s.

M.V. *Stentor*, Built 1946, GRT 10,203, NRT 6,053. Official Number 181024.

Sources: Francis E. Hyde, *Blue Funnel: a History of Alfred Holt & Company of Liverpool, 1865-1914* (Liverpool, 1957); S.W. Roskill, *A Merchant Fleet in War: Alfred Holt & Co, 1939-1945* (London, 1962).

## Appendix 3

### Kurt Hahn, 1886-1974

Kurt Hahn had a university education in Germany and at Oxford. He joined the German Foreign Office in 1914, becoming Private Secretary to the last Imperial Chancellor, Prince Max von Baden, at the end of World War I. With Prince Max he founded a co-educational boarding school at Salem in 1920. After persecution from the Nazis, he came to Britain in 1933 and founded a new public school, Gordonstoun, in 1934, becoming a British citizen in 1938.

His educational influence was vast and on an international scale. After World War II, the Outward Bound Trust (1946) was given the Aberdovey school by Holts, and numerous similar establishments were founded in Britain and overseas. Amongst other initiatives which may be traced back to Kurt Hahn were international schools commencing with Atlantic College in Wales, the training schooner *Captain Scott*, and the Duke of Edinburgh Award Scheme. Hahn provided the inspiration, then in most cases used his diplomacy and contacts to persuade others to carry them through.

Sources: D.A. Byatt, ed., *Kurt Hahn, 1886-1974: an Appreciation of his Life and Work* (Elgin, 1976); David James, ed., *Outward Bound* (London, 1957).

## Appendix 4

## S.V. *Passat*: dimensions and voyages

| | |
|---|---|
| Hull and Rig: | Steel four-mast barque |

| | | |
|---|---|---|
| Building/Owners | Blohm & Voss, Hamburg. | |
| | Cost, RM 680,000 | |
| | 1911-32 | Ferdinand Laeisz, Hamburg |
| | 1932-51 | Gustaf Erikson, Mariehamn |
| | 1951 | Belgian shipbreaker |
| | 1951-56 | Schliewen, Hamburg |
| | 1956-59 | Zerssen, Hamburg |
| | 1959-on | City of Lübeck/Travemünde |

| | |
|---|---|
| Tonnage | Gross 3,091. Net 2,882 (1911) |
| | Gross 3,181. Net 2,870 (1927) |
| | Gross 3,181. Net 2,593 (1951) |

| | |
|---|---|
| Dimensions | Length 322 ft. Breadth 48 ft. |
| | Depth 26.5ft. |

| | |
|---|---|
| Sails | Area 40,795 sq.ft. Royals, double topgallant, double top sails, courses; double spankers; staysails. Fore, main and mizzen masts identical, thus sails interchangeable. |

| | |
|---|---|
| Masts | Height above deck: fore 172 ft., main 165 ft., mizzen 172 ft., jigger 141 ft. Spike bowsprit 54 ft. Lower mast and topmast in one piece, jigger mast in one piece. |

| | |
|---|---|
| Engine | Fitted in 1951. 900 hp, six cylinder Krupp diesel giving about 6.5 knots under power alone. |

| | |
|---|---|
| Complement | Laeisz, about 36; Erikson about 30; Schliewen, Zerssen about 80. |

229

| Passages in 1952 | Brake to Rio Grande do Sul | |
|---|---|---|
| (author on board) | 10 February - 22 March | 41 days |
| | Rio Grande to Rosario | |
| | 10 - 16 April | 6 days |
| | Buenos Aires to Antwerp | |
| | 30 April - 16 June | 47 days |
| | Bremen to Santos | |
| | 3 July - 8 August | 37 days |
| | Santos to Diamante | |
| | 15 - 24 August | 9 days |
| | Buenos Aires to Antwerp | |
| | 27 Sept - 17 November | 52 days |

## Appendix 5

### List of Clothing Supplied by Schliewen & Co to
### *Passat* Cadets

Oilskin suit
pr. Sea boots
Pudel (woollen bobble cap)
Two blue working suits
Black uniform suit
Uniform cap
White cap
Knife with sheath

Sou'wester
pr. Sea boot socks
Polar-necked jumper
White working suit
Blue tie
Two white shirts
Brown jacket
Sea bag (self made)

Source: Author's diary for 1952.

## NOTES

1    Graham White, *Socialisation* (London, 1977), Chap. 1.

2    Bryan Nolan, 'A possible perspective on deprivations', in Peter H.
     Fricke, ed., *Seafarer and Community: Towards a Social Understand-
     ing of Seafaring* (London, 1973), 85-96.

3    See, Alston Kennerley, 'Seamen's missions and sailors' homes:
     spiritual and social welfare provision for seafarers in British ports
     in the nineteenth century, with some reference to the South West',
     in Stephen Fisher, ed., *Studies in British Privateering, Trading
     Enterprise and Seamen's Welfare, 1775-1900* (University of Exeter,
     1987), 121-65, Appendix 13.

4    Warren H. Hopwood, 'Some problems associated with the selec-
     tion and training of deck and engineer cadets in the British Mer-
     chant Navy', in Fricke, ed., *Seafarer and Community*, 97-116.

5    Tony Lane, *Liverpool. Gateway of Empire* (London, 1987), 98.

6    Lane, *Liverpool*, 96-101.

7    C. Fenton, *The Sea Apprentice: Being an Account of Two Voyages
     Round the World Under Canvas* (London, 1934), 5; A.H. Rasmussen,
     *Sea Fever* (London, 1952), Chap. 1.

8    Knut Weibust, *Deep-Sea Sailors: a Study in Maritime Ethnology*
     (Stockholm, 1969), 212.

9    White, *Socialisation*, 107.

10   Frank T. Bullen, *The Log of a Sea Waif* (London, 1899, 1910), 2; *Con-
     fessions of a Tradesman* (London, 1908), Chap. 3.

11   Jocelyn Fitzgerald Ruthven, *Memoirs of Jocelyn Fitzgerald Ruthven,
     Master Mariner, 1849-1943* (Ipswich, 1949), 3.

12   Geoffrey Rawson, *Sea Prelude* (London, 1958), 22-3.

13   Warren Hopwood, 'Training cadets in the Merchant Navy', in
     Fricke, ed., *Seafarer and Community*, 110-1.

14   Brian Nolan, 'A possible perspective on deprivations', in Fricke,
     ed., *Seafarer and Community*, 90-3.

15   Family episodes are based partly on the author's recollections
     and partly on unpublished autobiographical notes written by his
     mother, Mrs V.I. Pritchard (formerly Kennerley, née Wain-
     wright). The sea career of Captain W.G. Wainwright has been
     published in the *Newsletter of the Liverpool Nautical Research
     Society*. Subsequent research has revealed that Captain Wain-
     wright's father served as a master at sea, and that his uncle was
     also at sea.

16   For general particulars of Alfred Holt & Co, see Appendix 2.

17   Whether apprentice or cadet officers, or boy ratings, those that
     had spent a year aboard the training ships for officers (H.M.S.
     *Conway* or H.M.S. *Worcester*), or, before they had all closed, the

various training ships for boy seamen (for example the Marine Society's T.S. *Warspite* or the T.S. *Exmouth*), were exceptions. See Alston Kennerley, 'The education of the merchant seaman in the nineteenth century', (unpublished M.A. thesis, University of Exeter, 1978), Chaps 6 & 7.

18  Fenton, *Sea Apprentice*, 9.

19  See Kennerley, 'Education of a merchant seaman'. For a description of experience at a nautical college pre-sea course in the 1950s and a first voyage on a power-driven cadet ship see, Tony Lane, 'Neither officers nor gentlemen', *History Workshop Journal* (Spring, 1985), 128-43.

20  Kurt Hahn, 'The origins of the Outward Bound Trust', in David James, ed., *Outward Bound* (London, 1957), 1-17. See also Appendix 3.

21  Both ships had been built by Blohm and Voss, Hamburg, for Ferdinand Laeisz' 'Flying P' line of sailing ships, *Pamir* in 1905 and *Passat* in 1911, to a similar design. They passed to the ownership of Gustaf Erikson of Mariehamn between the wars, but were reacquired by German interests after the Second World War to reestablish sail training, sail experience still then being a requirement for officer qualifications. *Pamir* foundered in 1957, but *Passat* survives as a national monument, moored in Travemünde. For general particulars of *Passat*, see Appendix 4.

22  From 1 January 1952 the author kept a diary, from which he has been able to augment his memory of his experiences. Stan Hugill (1906-1992), went to sea at fourteen, and saw service in square rig and with Blue Funnel. He was Bosun at the Outward Bound Sea School in Aberdovey from 1950. His *Shanties from the Seven Seas* (London, 1961) is the definitive work on sea shanties.

23  White, *Socialisation*, 107.

24  Weibust, *Deep Sea Sailors*, 210-27.

25  See Appendix 5 for the author's cadet outfit.

26  Clifford W. Ashley, *The Ashley Book of Knots* (New York, 1944, Doubleday, 1956).

# NOTES ON CONTRIBUTORS

Margaret Deacon is a member of the Department of Oceanography, University of Southampton.

Harry N. Scheiber is Stefan Riesenfeld Professor of Law in the School of Law of the University of California at Berkeley, U.S.A.

Stella Maris Turk is an Honorary Research Fellow in the Cornish Biological Records Unit, of the Institute of Cornish Studies, University of Exeter.

A.J. Southward and Gerald Boalch are respectively Leverhulme Fellow and Honorary Research Fellow in the Marine Biological Association, Plymouth.

Michael Stammers is Keeper of the Merseyside Maritime Museum, Liverpool.

John F. Travis is an Occasional Lecturer in Social History in the Department of Economic and Social History, University of Exeter.

John Channon is a Senior Lecturer in the School of Slavonic and East European Studies, University of London.

Kim Montin is Curator of the Sjöhistoriska Museet at Åbo Academy University, Finland.

Alston Kennerley is a Principal Lecturer in the Institute of Marine Studies at the University of Plymouth.

Stephen Fisher is a Co-Director of the Centre for Maritime Historical Studies, University of Exeter.